The Quality of Ayurveda Education in India: A Survey

(Based on the doctoral thesis of Banaras Hindu University)

Kishor Patwardhan

Department of Kriya Sharir,
Faculty of Ayurveda, Institute of Medical Sciences,
Banaras Hindu University,
Varanasi, INDIA

Sangeeta Gehlot

Department of Kriya Sharir,
Faculty of Ayurveda, Institute of Medical Sciences,
Banaras Hindu University,
Varanasi, INDIA

H.C.S. Rathore

Faculty of Education,
Banaras Hindu University,
Varanasi, INDIA

Publisher: www.lulu.com, Raleigh, N.C. USA

The quality of Ayurveda education in India: A survey
(Based on the doctoral thesis submitted at Banaras Hindu University)

Publisher: www.lulu.com, Raleigh, N.C. USA

First Edition

Copyright © 2013, by Authors

All rights reserved. Without limiting the rights under copyright reserved above, no part of this publication may be reproduced, stored in or introduced into a retrieval system, or transmitted, in any form, or by any means (electronic, mechanical, photocopying, recording, or otherwise) without the prior written permission of the copyright owner of this book.

ISBN: 978-1-304-48764-3

Preface

The idea of engaging myself in this kind of work occurred to me for the first time when I took up the teaching profession as a lecturer. During my tenure as a lecturer, I realised that most of the students were disappointed with the quality of Ayurveda education in general. To my surprise, even many of the teachers were not satisfied with their job of teaching. The most commonly cited reason for this state of affairs was that 'a bare minimum stuff in the existing curriculum of Ayurveda courses was practical and clinically useful'. This was also the reason why most of the students would join some allopathic hospital as resident doctors for varying periods after completing their internship and would get themselves acquainted with some clinical skills required for general practice, and finally, would set up a clinic and practice allopathy. Surprisingly, no methodical study was available to me at that point of time that could have thrown some light on the condition of Ayurveda education. In fact, 'education' is never considered a potential area for research in Ayurveda even today.

This situation disturbed me and motivated me to carry out a methodical study that would provide me with some evidence to say that the quality of Ayurveda education was really poor. Hence, I prepared a preliminary questionnaire and collected some data from the students of a single institution and found that the results fairly matched with my presumptions. I presented this pilot study in the form a scientific paper at the international conference, RAAM (Recent Advances in Ayurvedic Medicine) held at Banaras Hindu University in 2003. At that point of time too, I faced a lot of criticism: most of the prominent academicians present there were in the mode of denial. They denied recognising the gravity of the findings stating that the particular institution where the study was carried out was probably inferior and the poor quality of education in that institution was not to be generalised. They even advised me not to present my findings on any platform in future. Later on, during the discussion, Prof. R. H. Singh suggested that the inclusion of teacher community would have made my study more credible as the students were supposedly immature.

Eventually, when I joined as a lecturer in the Faculty of Ayurveda at BHU, I thought of taking up this study once again in a planned way. This time, I got myself registered as a Ph.D. scholar and took up the study titled 'Relevance of current system of Ayurvedic education in the emerging global scenario' under the supervision of Dr. Sangeeta Gehlot. I approached Prof. H. C. S. Rathore for his help because the topic was related to his field of expertise, i.e., education. He stood by my side assuring me all the support that I needed.

I used a validated and methodically tested questionnaire as a tool which was mailed to about 32 Ayurvedic colleges spread across India. Selection of these institutions was based on the random cluster sampling method. The interns, the postgraduate students and the teachers from these institutions participated in the study. During the study too, I faced criticism; especially from the side of the teachers. Many of them were disturbed after reading some of the statements present in my questionnaire. This was probably because I was asking some fundamental questions in a very straight forward manner that posed some kind of threat to the work they had been doing all these years. Many of them refused to fill in the questionnaire. Some of them even felt that I was engaged in something that was against the 'Science of Ayurveda'. I must clarify that I had nothing against the 'Science of Ayurveda' in my mind; rather, I am an ardent fan of this ancient wisdom. However, I was, and still I am, of the view that the 'Education System in Ayurveda' is at fault.

Finally, my study revealed almost similar findings as were mentioned in my earlier paper that I presented at RAAM-2003, and my presumptions got further justified. However, I wish to state that I am not particularly happy with the results of this study. This is because, the study reveals a spectrum of lacunae in the existing system of Ayurveda education and this leaves me in a disturbed state of mind. Therefore, I decided to publish these findings in the form of scientific papers in scholarly journals. Three papers derived out of this work have been already published in peer reviewed journals. Though, publishing these findings will attract criticism, I thought, will also enable the policy makers to plan some concrete solutions to tackle these problems. With the same scientific spirit, I have decided to publish the gist of my thesis in the present form.

I would like to express my heart-felt gratitude to my supervisor **Dr. Sangeeta Geholt,** Associate Professor and Head, Department of Kriya Sharir, Faculty of Ayurveda, Banaras Hindu University, for having given her valuable time and guidance along with all possible support from the department through all these years. She has provided a conducive working atmosphere in the department which has had a positive impact on the current work. While writing the discussion, I have taken her help, especially when suggesting solutions to the current problems.

I take this opportunity to thank my co-supervisor, **Prof. H. C. S. Rathore,** Faculty of Education, Banaras Hindu University, for having provided his priceless inputs at all stages of the current work – in planning the study, in preparing the questionnaire and in drawing observations and inferences. His constructive and critical remarks have been vital in giving the present work this shape.

This work would not have been possible without the help that I received from **Dr. Girish Singh,** Scientific Assistant, Division of Bio- Statistics, Department of Community Medicine, Institute of Medical Sciences, Banaras Hindu University.

I would like to thank the following teachers and students who helped me in the collection of data from various institutions:

Prof. Abhimanyu Kumar, Prof. B. Sammaiya, Prof. D. N. Sharma, Prof. Dole, Prof. Jyoti Lal, Prof. M. S. Meena, Prof. Pradeep Bharadwaj, Prof. Sanjeev Rao, Prof. V. V. Chikara, Prof. Yeriswamy, Dr. A.K. Singh, Dr. Anil Dutta, Dr. Bhan Pratap Yadav, Dr. Brijesh Mishra, Dr. Chandra Shekhar Pandey, Dr. D. N. Mishra, Dr. Dinesh Kumar Malaviya, Dr. H. H. Awasthi, Dr. K. K. Dwivedi, Dr. Khagen Basumatary, Dr. Mahendra Prasad, Dr. Mangalagowri V Rao, Dr. Manjry Anshumala Barla, Dr. Manoj Kumar, Dr. Nagraj Poojary, Dr. Piyush Tripathi, Dr. Prakash Mangalasseri, Dr. Rahul Sukhdev Bankar, Dr. Rajni Sushma, Dr. Rakesh Roushan, Dr. Rakesh Tiwari, Dr. Rashmi Sharma, Dr. Sanjay Pokharel, Dr. Seema Joshi, Dr. Shrinidhi K.V., Dr. Sridurga, Dr. Sumer Singh, Dr. Supreet Joyal Lobo, Dr. Surya Prakash Mishra, Dr. Sushma Tiwari, Dr. Ved Prakash Vyas and Dr. Virendra Kumar.

I also thank all the Principals/ Deans/Directors of all the institutions covered in the study for having given their consent for the study to be carried out. My sincere thanks are due, to all my colleagues, junior & senior residents and non-teaching staff of the department of Kriya Sharir for extending their kind co-operation throughout the period of present work in all possible forms. I sincerely thank all those teachers and students who participated in the study by sparing their time. Many of them have offered some important suggestions and comments, which have been considered while writing the 'Discussion' part of the study.

Prof. Muralidhara Sharma, SDM College of Ayurveda, Udupi, Prof. P. N. Rao, SDM College of Ayurveda, Hassan and Prof. G. S. Acharya, SDM College of Ayurveda, Udupi have constantly been encouraging me in all my endeavours and I take this opportunity to thank them.

I thank my wife Medha for all the support she has extended during this work. I have taken her help while finalising the questionnaire and also while writing the thesis. She has also helped me in copy-editing the present work. Our son, Satvik and our daughter, Samyata have been the source of happiness and contentment in our lives and I thank both of them.

My parents, Prof. N. G. Patwardhan & Smt. Vijayalaxmi N. Patwardhan and my brother Vikram & his wife Archana have always been encouraging throughout my career and my special thanks are due to them.

Date: 04-11-2013
Varanasi   **Kishor Patwardhan**

Contents

Chapter	Page Number
Introduction	1
Ayurveda Education in Ancient India	5
Problems in the Current System of Ayurveda Education	17
The Study Design	28
Observations and Result	35
Discussion	80
Summary and conclusion	92
Bibliography and References	98
Annexure	104

Introduction

Ayurveda is the native Indian system of healthcare that is currently used by millions of people in India, Nepal and Sri Lanka for their day-to-day healthcare needs. *Vedas*, especially *Ṛgveda* and *Atharvaveda* (5000 years B.C.), the earliest documented ancient Indian knowledge have references on health and diseases. Ayurveda texts like *Caraka Saṁhitā* and *Suśruta Saṁhitā* were documented about 1000 years B.C. The term 'Ayurveda' means 'Science of Life'. This science deals elaborately with measures for healthy living during the entire span of one's life. Besides dealing with principles for maintenance of health, it also delves upon a range of therapeutic measures to fight diseases. Ayurveda was taught as a scientific subject in the oldest Indian Universities of Takshashila and Nalanda. The subject basically covered two schools – one of the physicians and the other of surgeons. It also had eight specialities. Standard Ayurveda books on these specialties were used in education for the training of students. The *'Gurukula'* system of education was the method of the Ayurveda training that was generally followed in ancient India. A 'Gurukula' was a place where a teacher or 'Guru' lived with his family and establishment, and trained the students. Caraka Samhita, one of the most popular textbooks on Ayurveda, delineates the process of selecting a suitable textbook and also an appropriate teacher by a disciple besides describing the three ideal methods of learning: self-study, teaching and discussions. The Gurukula system of education suffered a setback during the medieval and colonial periods of Indian history. Towards the end of British rule, the system received support from national leaders and its revival started. Separate schools of Ayurvedic education began springing up, first in several princely States and then followed by the State Governments of British India. The Indian National congress in 1920, adopted a resolution which strongly urged to popularize Ayurvedic schools, colleges and hospitals for promoting its education and practice in India. Following this, Ayurvedic schools and colleges were opened in *Varanasi, Madras, Delhi, Bombay, Bengal, Mysore* and in some other States of British India. As a result, the number of such institutions, which was 15 in 1920, increased to 30 in 1935 and reached a figure of 50 by the time the country achieved its independence.

In 1946, the conference of Health Ministers strongly recommended for starting of schools and colleges for diploma and degree courses in Ayurveda. It also recommended for the provision of Ayurvedic post-graduate courses for the graduates of western medicine. Another landmark in the education field was reached, when in 1946 Chopra Committee directed the educational process towards reaching the goal of achieving integration with western medical education system. Later, in 1949 Pandit Committee further strengthened this recommendation. The Chopra Committee suggested evolving a scheme of education wherein the teaching of Indian medicine would include the essentials of western medicine, particularly in those branches where Indian medicine was deficient. It also proposed that such bilateral instructions may be given till the ultimate object of integration leading to synthesis was achieved.

By the year 1958 there were about 75 institutions teaching Ayurveda. Out of these, some 50 institutions had adopted integrated pattern of education. Around this time the universities in the country started recognising the Ayurvedic institutions and seven integrated institutions got affiliated to universities. The academic control of the remaining teaching institutions was by the State Boards constituted by the State Governments.

With a view to streamline education and to evolve uniform standards for the Indian Systems of Medicine, the Government of India set up the Central Council of Indian Medicine through an Act of Parliament - the Indian Medicine Central Council Act, 1970. The Education Committee of this Central Council deals with all matters pertaining to the education of Ayurveda.

At present, the county has around 250 under-graduate colleges. More than 50 institutions have facilities for post-graduate education in different specialities of Ayurveda. The duration of under-graduate course is 5½ years after 10+2 Bio-Science education and the post graduate [*Ayurveda Vācaspati* - M.D.(Ayurveda) and *Ayurveda Dhanvantari*- M.S.(Ayurveda)] course is of further 3 years after graduation.

Ayurveda and the Current Global Scenario:

'Ayurveda' is generally considered as one of the Complementary and Alternative systems of medicine. Complementary medicine refers to a group of therapeutic and diagnostic disciplines that exist largely outside the institutions where conventional health care is taught and provided. In the 1970s and 1980s these disciplines were mainly provided as an alternative to conventional health care and hence became known collectively as "alternative medicine." The name "complementary medicine" developed as the two systems began to be used alongside (to "complement") each other. Over the years, "complementary" has changed from describing this relation to defining the group of disciplines itself. Some authorities use the term "unconventional medicine" synonymously (Zollman C and Vickers A, 1999).

WHO regards Ayurveda as one of the streams of Traditional Medicine (T.M.). It defines Traditional Medicine as follows:

"Traditional medicine includes diverse health practices, approaches, knowledge and beliefs incorporating plant, animal and/or mineral-based medicines, spiritual therapies, manual techniques and exercises, applied singularly or in combination to maintain well-being as well as to treat, diagnose or prevent illness."

During the period of last 30 to 40 years, several countries in the Asia have seen progress in incorporating their traditional health systems into the national policies. In some countries such as China and Vietnam, this development has been achieved by mobilizing all healthcare resources to meet national objectives for primary health care. In these countries, modern and traditional medicines are integrated through medical education and practice, representing an 'Integrated approach'. In countries like India and South Korea, a 'Parallel approach', where modern and traditional medicines are kept separate within the national health system is being followed (Bodeker G, 2001).

However, the Department of Indian Systems of Medicine has expressed concern over the substandard quality of education in many colleges, which, in the name of integration have produced fusion curricula and graduates, neither acceptable to modern nor to traditional standards.

Picture of Ayurveda in India and overseas has undergone a phenomenal change during last 20 to 25 years. A large population from all over the world is attracted towards this ancient system of healthcare because of the terms associated with it like 'Holistic Medicine', 'Herbal', 'Free from side-effects', 'Mind-Body and Spiritual approach' etc. Centres offering 'Panchakarma therapy', 'Ayurvedic Lifestyle Management' and 'Ayurvedic Massage' are being increasingly established. The use of Ayurvedic medicines has become accepted in other countries as well. For example, according to the 2007 National Health Interview Survey, more than 200,000 US adults had used Ayurvedic medicine in 2006 alone. Marketing strategies of major pharmaceutical firms have changed and 'Ayurvedic wings' of drug manufacturing have begun. In 2007, there were more than 8400 licensed Ayurvedic pharmacies in India and the approximate turnover of this industry was Rs. 4000 crore, which accounted for nearly a third of the total pharmaceutics business of the country (Thatte U, Bhalerao S, 2008). Many medical schools and other institutes all over the world have started offering some degree or diploma in Ayurveda. Several publication houses of international repute have started publishing literature related to Ayurveda. Even the people with no formal Ayurvedic education have started showing interest in authoring books and research papers on Ayurveda (Wujastyk D and Smith FM, 2008). Ayurveda is being seen as a rich resource for new drug development by modern day pharmacologists (Patwardhan B, Vaidya ADB and Chorghade M, 2004).

On the contrary, questions on safety and efficacy of Ayurvedic products are also being raised. In 2004 December, Journal of American Medical Association (JAMA) published a research paper which concluded that one of 5 Ayurvedic Herbal Medicine Products (HMPs) produced in South Asia and available in Boston South Asian grocery stores contained potentially harmful levels of lead, mercury, and/or arsenic. The paper also suggested that the users of Ayurvedic medicine may be at risk for heavy metal toxicity, and testing of Ayurvedic HMPs for toxic heavy metals should be made mandatory (Saper RB and others, 2004). This concern has led some countries like Canada to curb the import of Ayurvedic preparations from India. In 2005, the testing by Canadian Government revealed alarmingly high levels of heavy metals in the exported Ayurvedic medicinal preparations. The analysis highlighted the 'higher than acceptable concentrations' of heavy metals such as lead, mercury and arsenic. A similar paper appeared in JAMA in 2008 too, raising an alarm against the use of Ayurvedic products because of their possible heavy-metal contamination (Saper RB and others, 2008). National Policy on ISM and H, 2002 has also admitted that the safety, efficacy, quality of drugs and their rational use have not been assured in India. The said document states that there is no assurance whatsoever that formularies and pharmacopoeial standards are being followed by the Indian Systems of Medicine drug manufacturers.

Thus, Ayurveda is globally being perceived in several contradictory ways. Poor quality of Ayurveda graduates produced as a result of poorly structured and poorly regulated education system is at least one of the important factors responsible for this scenario. The number of Ayurveda colleges has increased phenomenally to 242, out of which, about 150 colleges have been established after 1980. Though the CCIM has implemented various educational regulations to ensure minimum standards of education, there has been a mushroom growth of sub-standard colleges causing erosion to the standards of education. Liberal permission by the State Governments, loopholes in the existing Acts and weakness in the implementation of standards of education have been held responsible for this state of affairs (National Policy on ISM&H, 2002).

Considering these facts, the present study was planned to evaluate the 'Relevance of current system of Ayurvedic education in the emerging global scenario'. The study is based on the perceptions of Ayurvedic students and Ayurvedic Teachers from various educational institutions spread all over India, on different aspects of Ayurvedic education.

Following were the Objectives of the study:

To evaluate the perceptions of Ayurveda students and Ayurveda teachers related to the relevance of the current system of Ayurveda education with special reference to the following aspects:
1. Exposure to the basic clinical skills during BAMS Course
2. Job opportunities after the completion of BAMS course
3. Scientific relevance of the Curriculum of BAMS course
4. Teaching methodology at graduate level Ayurvedic education.
5. Global Challenges being faced by the Ayurvedic system of education
6. Entrepreneurship /Business opportunities after the completion of BAMS
7. Ideal system of medical education for India

8. Personal relevance of Ayurveda to the individuals.

For this purpose, a validated and methodically tested questionnaire was prepared as a tool on the basis of interactions the investigator had with students and teachers of various educational institutions.

The final validated and tested questionnaire for its reliability was mailed to about 32 Ayurvedic colleges randomly, spread all over India so that at least 5 institutes were covered in each geographical zone (North, East, South and West). Selection of these institutions was based on the random cluster sampling method. The heads of the institutes were requested to distribute the questionnaire among Interns, Post Graduate students and Teachers. A period varying from 1 to 2 days was given to the respondents to return the filled questionnaires. The questionnaires thus filled were collected and the data was fed on to the computer as explained earlier. After the completion of the study, the results were analysed.

The participants were grouped under two categories: Students and Teachers. Tables of frequency and percentage were framed on the basis of responses to individual items with reference to the status of the respondents (Student or Teacher). Independent samples - T test was applied to compare mean scores of the two groups.

A total of 1022 participants from 18 States responded to the questionnaire. This number included 644 students and 378 teachers. The majority of participants (195) were from the state of Uttar Pradesh and the least number (19) of them were from the state of Bihar.

The results of the study indicate that a radical change is required at all layers of Ayurveda education at graduate level. Study indicates that exposure to basic clinical skills is inadequate and job opportunities are limited for BAMS graduates at present. Teaching methodology requires to be modified from literature- oriented teaching to skill-oriented learning. The study also suggests that some basic modifications are required in curriculum too. Some changes in the examination system too are required. The study indicates that students are not exposed to some essential technological advances making them technologically inferior.

Ayurveda Education in Ancient India

India has a rich tradition of teaching and learning right from the ancient times. It was the knowledge of acoustics *(Śruti)* that enabled ancient Indians to orally transmit the *Veda*s down the ages. Institutional form of imparting learning came into existence in the early centuries of the Christian era. Before this, the approach to learning was to study 'Logic' (the branch of philosophy that deals with the theory of deductive and inductive arguments and aims to distinguish good from bad reasoning) and 'Epistemology' (the branch of philosophy that studies the nature of knowledge, in particular its foundations, scope, and validity). One of the most important topics of Indian thoughts was *Pramāṇa* or means of acquiring the reliable knowledge. The *Nyāya* School advocated four *Pramāṇas* – *Pratyakṣa-* direct perception, *Anumāna-* inference drawn by applying the logical analysis, *Upamāna-* understanding by analogy or comparison and *Śabda-* pronunciation of a reliable authority such as an authoritative textbook.

Bhatta CP observes that in ancient India, the thinking principle was regarded to be higher than the subject of thinking. So, the primary subject of education was the mind itself. According to the ancient Indian theory of education, the training of the mind and the process of thinking are essential for the acquisition of knowledge. So, the pupil had mainly to educate himself/herself and achieve his/her own mental growth (Bhatta CP, 2007). According to Rangachar S, the direct aim of ancient Indian education was to make the student fit to become a useful and pious member of society (Rangachar S, 1964). RK Mookerjee opines that inculcating the civic and social duties among the students was also a part of ancient Indian educational system. The students were not to lead a self-centred life. They were constantly reminded of their obligations to the society. Convocation address to the students as found in *Upaniṣads* show how they were inspired to be useful members of the society (Mookerje RK, 1989). Markandan N states that ancient Indian educational system focussed on building a disciplined and value-based culture. Human values such as trust, respect, honesty, dignity, and courtesy were given importance in the education system (Markandan N, 2001).

The convocation address found in *Taittirīya Upaniṣad* throws significant light on the qualities required to be developed in the students which are not very different from the qualities that modern educational systems are trying to impart.

"Speak the truth. Practise righteousness. Make no mistake about study. There should be no inadvertence about truth. There should be no deviation from righteous activity. There should be no error about protection of yourself. Do not neglect auspicious activities. Do not be careless about learning and teaching. There should be no error in the duties towards the gods and manes. Let your mother be a goddess unto you. Let your father be a god unto you. Let your teacher be a god unto you. Let your guest be a God unto you. The works that are not blameworthy are to be resorted to, but not the others. The offering should be with honour; the offering should be in plenty. The offering should be with modesty. The offering should be with sympathy. Then, should you have any doubt with regard to duties or customs, you should behave in those matters just as the wise men do, who may happen to be there and who are able deliberators, who are adepts in those duties and customs, who are not directed by others, who are not cruel, and who are desirous of merit. This is the injunction. This is the instruction. This is the secret of the scriptures." (Taittirīya Upaniṣad, I. xi.1-4).

This convocation address outlines some of the domestic and social duties of students in very clear terms. Accordingly students are to honour father, mother, teacher and guest as gods; to honour superiors; to give in proper manner and spirit, in joy and humility, in fear and compassion. Lastly, the pupil is also asked not to neglect his health and possessions. This convocation address is very important in understanding the role of ancient Indian education in building a values-based culture (Ramajois M, 1987).

Selecting a Textbook:

Caraka Saṁhitā gives details related to the methods followed in Medical education.

[Most of the text in this section has been adopted as such from the English translation of Caraka Samhita by Sharma R.K. and Dash Bhagwan (2006).]

A wise person, desirous of adopting medical profession should, first of all, carefully select a suitable text on medicine, depending upon his competence to undertake light or serious type of work, his willingness for short term or long term results, his habitat and age. There are several such texts available for physicians. Only the texts having the following characteristic features are to be followed:

1. Which are followed by great, illustrious and wise physicians;
2. Which are charged with ideas and respected by reputed experts;
3. Which are conducive to the intellectual growth of disciples of all the three categories (viz. highly intelligent, moderately intelligent and less intelligent);
4. Which are free from defects of repetition, which are transmitted by seers and have well-knit aphorisms together with commentaries thereon in proper order;
5. Which have elegant ideal to convey;
6. Which are free from vulgar and difficult expressions and have clear and unambiguous expressions;
7. Which convey ideas in an orderly manner;
8. Which primarily deal with the determination of real objects;
9. Which are free from contradictions
10. Where there is no confusion relating to contexts;
11. Which convey ideas quickly; and
12. Which are equipped with definitions and illustrations. A text of this type may be compared to the Sun which removes darkness and illuminates all (Ca.Vi.8/3).

This explanation implies that the student willing to become a physician had to have a prior basic understanding related to the medical profession in general and had to go through the available popular textbooks before choosing the one that best suited him or her depending on the personal preferences. This was because the medical education was based on a particular textbook related to a particular speciality in ancient India. This means that at the initial stage of medical education itself, the student had to have a clear goal related to his profession.

In the present-day education, the scenario has changed. After completing the undergraduate education, during which a gross idea related to each subject speciality is given, the student chooses the field of specialization only at the postgraduate level. Even the prior knowledge related to the respective medical field is not considered essential today, especially in Ayurvedic education. During primary, secondary and pre-university level education, the students are exposed to the basics of human anatomy, physiology and to some extent medicine in terms of contemporary medical science. No ancient knowledge related to Ayurveda is introduced at any level of education in India today. Probably this situation has created some problems in Ayurvedic education because the students getting themselves admitted in the

Ayurvedic undergraduate programmes generally have no basic idea related to Ayurveda. The solution for this problem could be to introduce some basic concepts of Ayurveda at primary level of education itself.

Selection of a Teacher:

In Caraka Saṁhitā, the process of selecting a suitable preceptor or Guru is described as follows:

> "One should assess the qualities of the preceptor. An ideal preceptor is one who is well grounded in scriptures; equipped with practical knowledge, wise, skilful, whose prescriptions are infallible, who is pious, who has all the necessary equipments for treatment, who is not deficient in respect of any of the sense organs, who is acquainted with human nature, and the rationale of treatment, whose knowledge is not overshadowed (by the knowledge of other scriptures), who is free from vanity, envy and anger, who is hard working, who is affectionately disposed towards his disciples and is capable of expressing his views with clarity"(Ca.Vi.8/4). Caraka further explains that "One should approach such a preceptor and respect him like fire, God, king, father and master with all care. After having obtained the knowledge of the entire scripture, through his blessings, one should strive again and again for achieving depth in the scriptures, clarity of expressions and comprehension of the various concepts and power of oration"(Ca.Vi.8/5).

This narrative clearly describes the liberty a student had in selecting his or her teacher. This situation exists even today in a modified form. Today, the student has a liberty to choose the institution of his or her choice based on his/ her performance in the pre-admission tests conducted at various universities or colleges. A single teacher of ancient times is represented by a whole institution in the present scenario of education.

Methods of Study:

In Caraka Saṁhitā, the methods of study have been described as follows: (Ca.Vi.8/6)
(1) Self- study -*Adhyayna*
(2) Teaching –*Adhyāpana* and
(3) Participation in debates and discussions – *Tadvidyāsaṁbhāṣā*

Procedure for self-study:

> "The disciple should be healthy and solely devoted to study. He should get up early in the morning or in the last quarter of the night. He should then perform ablution and offer prayers to the gods, sages, cows, *Brāhmaṇas*, teachers, elderly and enlightened persons and preceptors and should then sit comfortably on an even and clean place. Thereafter, he should recite the *Sūtras* orally with due concentration. After proper understanding, he should repeat his recitation with a view to remove his own deficiencies and testifying to the deficiencies of others. He should continue with his practice in the noon, in the after-noon and at night without any break. This is the procedure for self- study" (Ca.Vi.8/7).

Procedure for teaching:

> The preceptor planning to undertake teaching should, first of all, examine the disciple. **Qualities of a good disciple** have been enumerated as follows:
> 1. Serenity
> 2. Kindness
> 3. Aversion to mean acts
> 4. Normal eyes, face and nasal ridge
> 5. Thin, red and clear tongue
> 6. Absence of any morbidity in teeth, lips and voice
> 7. Determination

8. Freedom from pride
9. Presence of intellect, power of reasoning and memory
10. Broad-mindedness
11. Birth in the family of a physician or the one having the temperament of a physician
12. Inquisitiveness for truth
13. Physical perfection
14. Unimpaired senses
15. Modesty and absence of ego
16. Ability to understand the real meaning of things
17. Absence of irritability
18. Absence of addictions
19. Good character, purity, conduct, love for study, enthusiasm
20. Devotion to study
21. Uninterrupted taste for the theory and practice of the science
22. Absence of greed and laziness
23. Good-will for living beings
24. Obedience to all the instructions of the preceptor and
25. Devotion to the preceptor (Ca.Vi.8/8)

This explanation is interesting because this is indicative of the concept of a qualifying test that probably existed in ancient times. A student had to probably undergo some qualifying tests before undertaking the study and probably the teacher himself used to conduct these tests. These tests included the mental ability tests and medical examination to rule out any hormonal/ heritable problems possibly hindering the abilities to perceive knowledge. For example, a thick tongue may be indicative of hyperpituitarism whereas a flat nasal bridge may indicate some kind of Chondrodysplasia. These tests probably also included the assessment of emotional stability/ sensitivity/aptitude/humanitarian values and other similar spheres of personality.

This situation has changed today. Most of the qualifying tests today are only aimed at assessing the understanding of the students in the subjects like physics, chemistry and biology. No methods are adopted to assess the other facets of one's personality like emotional stability/sensitivity/aptitude/value system. Medical examination is conducted even today with the similar purpose but the methods have changed.

Procedure for Induction of a student:

In Caraka Saṁhitā, the following description is found in relation to the procedure for induction: When the disciple having the above mentioned qualities approaches the preceptor with reverence for study, he should be inducted as follows:

The physician should get constructed a *Sthaṇḍila*, (an elevated place of the shape of a square and of four cubits in size) in an even and clean place having slope towards the east or the north. The place should be smeared with cow dung, spread with *Kuśa* grass and provided with good border in all the four sides. This place should then be decorated with sandal paste, earthen jar, water, silken garments, ornaments of gold, silver, jewels, pearls, corals, food articles which promote intellect, fragrant things, white flowers, fried paddy, mustard seeds, and unbroken de-husked rice. Then fire should be ignited in that place with the help of dried twigs of *Palasha, Ingudī, Udumbara* or *Madhūka*. The physician facing towards the east with purity of mind and following the procedure of study, should offer oblations of honey and ghee in fire reciting benedictory *Mantra*s ending with *Svāhā* for *Brahmā, Agni, Dhanvantari, Prajāpati*, the *Aśvins, Indra, Ṛṣis* and authors of hymns -three times each.

The disciple should follow the preceptor. After offering oblations he should take a round of the fire keeping it to the right side. After taking the round, *Brāhmaṇas* may be made to recite propitiatory hymns. He should then offer prayers to the physicians (Ca.Vi.8/9-12).

In front of the fire, *Brāhmaṇas* and physicians, the preceptor should instruct his disciples as below:

1. You should observe *Brahmacarya*, maintain your beard, speak the truth, take vegetarian food, resort to such food and regimens as are conducive to the promotion of intellect, refrain from envy and carry no weapon with you.
2. You should always obey the instructions of your Guru except when they go against the ruler of the land, or they are directed towards your death or they involve considerably sinful commitments or bring about calamity.
3. You should always be devoted to your preceptor, surrender yourself to his superiority, be subordinate to him and behave in a manner which will be pleasant and useful to him.
4. You should pay due regard to your Guru as if you are his son, servant or supplicant.
5. You should act without ego, with care and affection with undisturbed mind, with modesty, with proper vigilance, without jealousy and with obedience for the instructions by your Guru.
6. Acting either at his instance or otherwise, you should first of all try to collect to the best of your ability the things desired by your preceptor.
7. If you want to achieve success in your medical profession, earn wealth as well as fame and attain heaven after death, you should in all circumstances pay for the well-being of cows, *Brāhmaṇas*, and all other living beings.
8. You should make efforts to cure the patient.
9. You must never give way to any ill will towards your patients even at the cost of your life.
10. You should not even think of committing adultery and should not aspire for any property belonging to others.
11. Your appearance and apparel should make you look modest.
12. You should not take wine, commit sins or have association with those committing sinful acts.
13. Your speech should be pleasant, pure, righteous, blissful, excellent, truthful, useful and moderate.
14. Your behaviour should be in conformity with the time and place, based on the recollections of the past experience.
15. You should always make efforts for the upliftment of your knowledge and adoption of such methods as would give you good health.
16. You should not prescribe medicines for those who are despised by the king or noble persons and those who despise the King or noble persons.
17. You should not treat all those who are excessively artificial in their behaviour, are wicked or are of miserable conduct and behaviour or who have not been absolved of the allegations against them or who are going to succumb to death.
18. Women in the absence of their husbands and guardians should not be treated by you.
19. You should not accept any enjoyable thing given by a woman without the permission of her husband or guardian.

20. You should enter the residence of the patient accompanied by a person who knows the place and who on his part, has obtained permission to enter there. While doing so, you should be well clad, with your head bowed down, having a good memory, having concentration of mind and acting with proper thinking. After having entered there your speech, mind, intellect and senses should be entirely devoted to nothing except the welfare of the patient and allied matters.
21. Family customs (secrets) should not be disclosed by you to outsiders.
22. Even having known that the patient's span of life has come to a close, you should not disclose this to the patient himself or to the son or father etc. of the patient because it may cause shock to the patient or to his relatives.
23. Even though actually possessed of wisdom, you should not exhibit it to others. Many people get very much irritated to hear such self-praise even from a saint (Ca.Vi.8/13).

It is not easy to acquire comprehensive knowledge of Ayurveda. Therefore, one should make honest efforts to be in constant touch with this science. One should strive to acquire qualities; one should learn similar noble qualities even from his enemies without having any sense of jealousy. The wise consider the entire universe as their preceptor; it is only the unwise who consider it to be their enemy. One should, therefore, have the proper advice which brings fame, which promotes longevity and nourishment and which is acceptable to the people. Such advice can be had even from an enemy and be adopted in practice.

Thereafter, the preceptor should say, "I will always behave well with the gods, fire, *Brāhmaṇas*, preceptors, elders, persons who have attained perfection and teachers. If you do so, fire, all types of smells, taste, jewels, seeds and the Gods will bless you. Otherwise, they will be unfavourably disposed towards you."

To the preceptor advising as above, the disciple should say, *"Tathā"*, i.e., "I shall act accordingly". It is only when the disciple acts accordingly; he can be considered eligible for studies. A teacher gets all the auspicious fruits of teaching those described in scriptures and even those that are not described and gets himself and the disciple endowed with the virtuous qualities only when the disciple is worthy of teaching (Ca.Vi.8/14).

In Suśruta Saṁhitā, the method of Initiation of pupil is described as follows:

The physician should initiate a student whose age is proper and suitable for study. The ideal student should have qualities like chastity, bravery, cleanliness, right conduct, politeness, strength, prowess, intelligence, courage, memory, wisdom and ability to grasp the meaning of words and interpret them. Also, the ideal student should have some physical qualities like thin lips, tongue and teeth; straight mouth, eyes and nose. Student's mind, speech and activities must be pleasant and he should be capable of withstanding strain. The teacher should not initiate any person who has qualities opposite of the above (Su.Sū.2/3).

On a day having auspicious stellar constellation and at a precious time of that day, the preceptor who is initiating the pupil, should select a place in an auspicious direction, which is clean and even; which is four arms square, smear it with cow dung and spread *Darbha* grass over it; worship Gods, *Brāhmaṇas* and physicians offering them gems, flowers, *Lāja* (fried paddy) and *Bhakta* (cooked rice). Then sprinkle holy water on the prepared ground, write the figure of Brahma on the right side and that of Agni nearby; then build a fire-altar and kindle the fire. Next, using twigs of *Khadira*, *Palāśa*, *Devadāru* and *Bilva* trees or of four trees with milky sap (such as *Nyagrodha*, *Udumbara*, *Aśvattha* and *Madhūka*), dipped in curds, honey and ghee; perform fire sacrifice in *Dārvī Homa* method, along with chanting of *Praṇava* (the term OM) and *Mahāvyāhṛti*, offering oblations of ghee to each God and sages, uttering the word *"Svāhā"* for each; the pupil should also be instructed to offer oblations of ghee to gods and sages (Su.Sū.2/4).

Brāhmaṇa (initiator) is eligible to initiate the pupils of all the three castes (*Brāhmaṇa, Kṣatriya* and *Vaiśya*); *Rājanya* (priest of warrior caste) for only two castes and a *Vaiśhya* (priest of merchant caste) for only pupils of *Vaiśhya* (merchant caste). Even a pupil belonging to a *Shūdra* caste, if possessing good qualities (knowledge, behaviour etc.) may also be initiated: so say some authorities (Su.Sū.2/5).

Next, going round the fire-altar three times, and in the presence of fire, instruct the pupils as follows:

"You should remain here, avoiding/foregoing desires, anger, greed, infatuation, pride, egoism, jealousy, harsh speech, finding fault in the speech of others, speaking untruth, laziness and acts of ill repute; keep yourself clean by cutting undesirable nails and hairs; put on ochre coloured clothes, maintain truthfulness; cultivate celibacy and habit of prostrating (to gods, preceptors, elders etc.) essentially; go only to such places, use only such bed, seat (for sitting) food and the mode of study which are approved by me; indulge in such activities which are liked by me and beneficial to me; doing any thing other than these, will be unrighteous on your part, your knowledge becomes futile and will not earn reputation" (Su.Sū.2/6).

"If on the other hand, I misunderstand you though you are behaving properly and act otherwise (punish, not teach well etc.) then, I will also acquire sin and my knowledge becomes futile" (Su.Sū.2/7).

"The twice born (*Brāhmaṇa, Kṣatriya* and *Vaiśya* castes), the preceptor, the poor, the friend, the ascetic, the refugee, the pious, the orphan, and the person coming from a distant place seeking help should be looked after like your own kin and treated with your own medicines, this is the ideal. The hunter, the bird catcher, the unrighteous and the sinner need not be cared for. By these acts your knowledge shines, you will acquire friends, fame, virtue, wealth and pleasures" (Su.Sū.2/8).

Procedure for Debate:

Caraka Saṁhitā gives following description related to the procedure for debate:

A physician should participate in a discussion with another physician. Professional discussion indeed promotes the power of application of knowledge and competition leading to enlightenment. It manifests the clarity of knowledge, promotes the power of speech, spreads fame, eliminates doubts reminiscent of the previous study by repeated hearing and brings about confirmation on of what is undoubtedly understood before. During the course of discussions one comes to know of many new things which were not heard by him previously. Being pleased over the devoted disciple, the preceptor during the course of teaching elaborates some secret meanings. The participants during the course of mutual discussion enthusiastically disclose these secret meanings in brief with a view to achieve a victory over the competitor. Therefore, participation in professional debates is always applauded by the wise.

Professional discussions are of two types, viz. (i) friendly discussions and (ii) hostile discussion (Ca.Vi.8/15-16).

Procedure for friendly discussion:

One should have friendly discussions with persons of learning possessed of scientific knowledge, power of argument and counterargument, who do not get irritated, who are endowed with correct knowledge, who are not jealous, who can be made to understand, who are competent in convincing others, who are capable of facing difficult situations and who can address in a sweet tone.

One should confidently discuss with such persons and pose questions to them. When he asks anything, it should be elaborately described with confidence. One should not get worried under the apprehension of getting defeated. One should not rejoice by defeating his opponents. One should not boast of having defeated such opponents. One should not hold extreme views under delusion. One should not try to describe a thing which the other party does not know. One should try to bring round the other party with politeness and not by deception. One should be very careful to behave politely with his opponents. This is the procedure for "Friendly discussions" (Ca.Vi.8/17).

Procedure to be adopted in a hostile discussion:

With persons other than preceptor and *Brahmacārins* (class-mates), one should go in for a "hostile discussions", provided he is confident of his superiority. Before entering into the discussion, the procedure proposed to be adopted by the opponent, difference between the abilities of himself and the opponent and the disposition of the members of the assembly should be carefully examined. A wise person determines the time of entering into or giving up the discussion only by proper examination. Hence proper examination is always advisable.

There are some good and bad qualities of the participants in a discussion. With a view to determine the superiority or inferiority of himself in respect of his opponent, one should carefully examine these good and bad qualities. Good qualities of participants are the knowledge of the text, practical experience, retention power, presence of mind and eloquence. Bad qualities of the participants are irritation, lack of skill, cowardice, lack of the power of retention and carelessness. One should compare the strength or weakness of himself and of his opponent in respect of these qualities (Ca.Vi.8/18).

Three types of opponents:

Depending upon the presence of the above mentioned qualities, the opponent may belong to any of the three categories, viz. (i) Superior, (ii) Inferior or (iii) Equal. However, other factors like the family status, conduct, religions etc. should not be taken into account in this connection (Ca.Vi.8/19).

Two types of assembly:

An assembly may be of two types, viz. (i) Enlightened and (ii) Dull. On the basis of different criteria, both these types of an assembly may be classified into three types viz. (i) Friendly, (ii) Neutral and (iii) Prejudicial. Members of an assembly may be enlightened (endowed with knowledge, experience, power of speech and contradiction) or dull but if they are prejudicial, then one should never enter into a discussion with anybody, not even with the most wretched one in such an assembly. If the members of the assembly are dull but friendly or neutral, then the individual should enter into discussion with an opponent who is not very famous and who is even despised by great people and who is without theoretical and practical knowledge or power. While discussing with such an opponent, one should use such long sentences as are difficult to understand or are composed of long and complicated aphorisms. An overexcited opponent should be ridiculed and the individual should continue his speech acting as if addressing the assembly, without giving an opportunity for the opponent to speak. One should speak using such terms as are difficult to understand and the opponent should be told that he was incapable of advancing any argument in the matter and his proposition had failed. If the opponent challenges again, he should be told, "You should study for at least one year more to have some more experience in debates. Probably you have not observed the guidance of your preceptor well." Or else, he should be told, "This is sufficient for you." Once the opponent is defeated he remains defeated forever; hence his further challenge for discussion should not be accepted. Some people advise that the same procedure should be followed even while discussing with a superior opponent. But the wise do not approve of such a proposition to enter into hostile discussion with a superior opponent. While discussing with the opponent in a debate, members of which are not enlightened, one should use complicated sentences so that members of the assembly will find it very difficult to understand. By implication, the user of such words and sentences will be credited with success. (Ca.Vi.8/20)

Procedure for debate with an opponent of inferior or equal type:

With an opponent of inferior or equal type, one should enter into hostile discussion if the members of the assembly are favourably disposed towards him. In an assembly where members are neutral and are attentive, learned, experienced, inclined to hear, having the power of retention, speech and contradiction; one should carefully observe the good and bad qualities of the opponent as a participant in the discussion. On the basis of these observations, if the opponent is found to be belonging to the superior category, then one should not enter into discussion on the same topic. Without letting the assembly know, he should change the topic of discussion to a favourable one. If the opponent is found to be of inferior category then efforts should be made to defeat him immediately in a hostile discussion. The following procedure should be adopted for immediately defeating an opponent of inferior category. If the opponent is not a learned person, then he should be defeated by citing long aphorisms; if he is not experienced then by such words and sentences as are difficult to understand; if he is unable to retain sentences by memory then by sentences composed of complicated and long aphorisms; if he is dull, then by statements of the same type (composed of the same words) but carrying different meanings; if he is devoid of the power of oration then by challenging with half of a sentence (the opponent in that case is required to fill up the other half); if he has no experience of participating in seminars, then by putting him to a disgraceful situation; if he is irritable, then by creating difficult situations for him; if he is a cowardice then by creating fearful situations and if he is not careful, then by adhering to the discipline of discussion. These are the procedures to be followed for immediately defeating the opponent of inferior category (Ca.Vi.8/21).

In fighting discussions one should make careful statements and should not over-rule the statements (of opponents) which are well authenticated. Some people get excessively irritated during hostile discussions and there is nothing which cannot be done or said by the enraged one. Therefore, in an assembly of learned people, the wise never appreciate a quarrel. This is how one should participate in a debate (Ca.Vi.8/22-24).

In the beginning one should proceed like this: One should prevail upon the assembly to select such a topic as is favourable to him and is exceedingly difficult for the opponent to discuss. Or the opponent should be made to take such a side in the discussion which will be disliked by the members of the assembly. When the assembly supports the stand taken by him, he should say, "I have nothing more to say. The assembly according to choice may sufficiently and appropriately decide upon the validity of the debate and its limitations," he should observe silence, thereafter (Ca.Vi.8/25).

The following factors bear importance in determining the limits of a fighting debate: (Ca.Vi.8/26)

1. Things which should be said;
2. Things which should not be said; and
3. The point of defeat.

Acquaintance with the following terms helps in the determination of the course of debate among physicians:

(1) *Vāda* (Debate), (2) *Dravya* (Substance), (3) *Guṇa* (Attributes), (4) *Karma* (Action), (5) *Samānya* (Generic concomitance), (6) *Viśeṣa* (Variant factor), (7) *Samavāya* (Inseparable concomitance), (8) *Pratijñā* (Proposition): (9) *Sthāpanā* (Justification), (10) *Pratiṣṭhāpana* (Counter argument), (11) *Hetu* (Cause), (12) *Dṛṣṭānta* (Example), (13) *Upanaya* (Subsumptive correlation), (14) *Nigamana* (Final conclusion), (15) *Uttara* (Rejoinder), (16) *Siddhānta* (Concluded truth), (17) *Śabda* (Words), (18) *Pratyakṣa* (Direct observation), (19) *Anumāna* (Inference) (20) *Aitihya* (Words of divine origin), (21) *Aupamya* (Analogy), (22) *Saṁśaya* (Doubt), (23) *Prayojana* (Object), (24) *Savyabhicāra* (Statements with exceptions), (25) *Jijñāsā* (Enquiry), (26) *Vyavasāya* (Determination) (27) *Arthāpatti* (Implied meaning), (28) *Sambhava* (Source), (29) *Ananuyojya* (Defective statement), (30) *Anuyojya* (Infallible statement), (31) *Anuyoga* (Scriptural enquiry), (32) *Pratyanuyoga* (Scriptural counter enquiry), (33) *Vākyadoṣa* (Syntactical defects), (34) *Vākyapraśaṁsā* (Syntactical excellence), (35) *Chala* (Casuistry), (36) *Ahetu* (Casual fallacy), (37) *Atītakāla* (Defiance of temporal order), (38) *Upalambha* (Pointing out defects in casuistry), (39) *Parihāra* (Correction), (40) *Pratijñāhāni* (Shift from the original proposition), (41) *Abhyanujñā* (Confessional retort), (42) *Hetvantara* (Fallacy of reason), (43) *Arthāntara* (Irrelevant statement) and (44) *Nigrahasthāna* (Clinchers) (Ca.Vi.8/27).

Improper time for study:

According to Suśruta Saṁhitā, the eighth day of the dark fortnight, the last two days (fourteenth and fifteenth) before its expiry; similarly of the bright fortnight; during sunrise and sunset, days of untimely (unseasonal) lightning and thunder, time of calamity of ones own place, country and the king; at the burial ground, time of riding (on animals or vehicles), at places of slaughter; during war, great festivals, and times of appearance of natural calamities (malevolent stars, comets etc.) and such other days on which the *Brāhmaṇas* do not study and at times when the person is not clean; on these days and times, studying of the science should not be undertaken (Su.Sū.2/9-10).

Importance of interpretation:

In Suśruta Saṁhitā, the following description is found in relation to the importance of interpretation:

Even after the science has been studied, if it is not properly interpreted in its meanings (implications) then it will only be causing exertion (to the person) just like the donkey carrying a load of sandal wood. Just as a donkey carrying a bundle of sandal wood understands only its weight and not the sweet smell of sandal wood, a foolish person, though having read many sciences, acts like a donkey in interpreting the meanings (implications) of the science.

Hence every word, part of the verse and the full verse of all the chapters should be explained (by the teacher) and understood (by the students). This is because the analysis of

Dravya (substances, drugs), their *Rasas* (taste/ chemical nature), *Guṇas* (qualities), *Vīrya* (potency), *Vipāka* (effects after metabolism), *Doṣas* (homeostatic mechanisms), *Dhātus* (tissues), *Malas* (wastes), *Āśayas* (viscera), *Marma* (vital spots), *Sirā* (veins), *Snāyu* (ligaments), *Sandhi* (joints), *Asthi* (bones), products forming the *Garbha* (embryo) and their combination, removal of foreign bodies lost inside the body, determining the nature of wounds, kinds of fractures, curability and fatality of diseases and such other topics are subtle and though analysed for a thousand times, still create confusion in the minds of even those who possess clear and wide knowledge; then what to say of the less intelligent? So, every word, part of a verse and the complete verse should be explained well by the teacher and made to understand by the students (Su.Sū.4/3-5).

Importance of exposure to other streams of sciences:

In Suśruta Saṁhitā, the following description is found in relation to the importance of exposure to other streams of science:

Meanings (implications) of even other sciences which are mentioned here (briefly during the course of teaching) should also be learnt by listening to the lectures of persons learned in those sciences, because it is not possible to include the details of all the sciences, in anyone science itself.

He, who studies only a single science, will not be able to arrive at a correct decision; hence the physician should have the knowledge of many sciences. He who practices medicine after learning the science from a preceptor and engaged constantly in recapitulating it, can only be called a physician, whereas all others are impostors (Su.Sū.4/7-8).

Style of Documentation:

The classical textbooks of Ayurveda are the documents of various observations by ancient physicians and surgeons. Regarding the style of documentation, with special reference to the sequence of arrangement of various topics, the following interesting facts can be short-listed:

- Initial set of chapters termed *'Sūtra Sthāna'* contains the summarised knowledge documented in the whole textbook. It covers a vast range of topics from basic sciences like anatomy and physiology to specialized areas like pharmaceutics and management of various diseases. Therefore, in the initial phase itself, the students were exposed to the complete nature of the science so that they could develop interest in any particular field as the study advanced.

- Next set of chapters termed *'Nidāna Sthāna'* contains basic pathophysiology of major sets of diseases in full – length. It is interesting to note that before teaching the detailed anatomy and physiology, the students were directly exposed to the topics related to pathology.

- After the description of pathology, the next set of chapters in Caraka Samhitā is *'Vimāna Sthāna'* whereas in Suśruta Samhitā, it is the *'Śārīra Sthāna'*.

- *Vimāna Sthāna* contains detailed descriptions related to means of deriving knowledge with special reference to diagnosis. Details of the clinical methods to be followed in diagnostics are described in this section. Some details of physiological and pathological features of different systems *(Srotāmsi)* in the body also are found here. Classification of diseases also has found its mentioning. Description of individuals belonging to different body constitutions have been mentioned in this section.

- *'Śārīra Sthāna'* contains some spiritual aspects like *Ātmā* along with anatomy, physiology including embryology, neonatology and obstetrics in both Caraka and Suśruta Samhitās.

- Next set of chapters in Caraka Samhitā is called *'Indriya Sthāna'*. This set is oriented mainly towards prognosis and features of imminent death.
- In Suśruta Samhitā, *Vimāna Sthāna* and *Indriya Sthāna* are not present.
- *'Cikitsā Sthāna'* in both Caraka and Suśruta Samhitās, contains the description of various methods of management of different sets of diseases. Interesting fact that draws attention in this regard is that the nomenclature of a disease in Ayurveda is mainly based on the prominent sign or symptom. For example, all diseases manifesting with fever as the presenting symptom are included under *'Jvara'* and all diseases manifesting with bleeding as the presenting symptom are classified as *'Raktapitta'*. In Caraka Samhitā, this section contains two chapters specially designated for *Rasāyana* (Geriatrics and rejuvenative medicine) and *Vājīkaraṇa* (Sexual Medicine).
- *'Kalpa Sthāna'* of Suśruta Samhitā contains details of toxicology whereas in Caraka Samhitā, it contains the details related to the various combinations of herbs used for *Vamana* and *Virecana* procedures. The next section in Caraka Samhitā, the *'Siddhi Sthāna'*, contains the details of procedures to be followed while administering *Pancakarma* therapy, especially, *Basti*. Suśruta Samhitā does not contain *'Siddhi Sthāna'*.
- In Suśruta Samhitā, there is another set of chapters called *'Uttara Tantra'*. This contains the details of diseases and treatment methods related to eye, ear, nose and throat.
- 'Aṣṭānga Hṛdaya' of Vāgbhata is a sort of compilation work which incorporates the major aspects from both Caraka and Suśruta Samhitās.
- Most of the other textbooks, which were composed later, are subject-specific. For example, *Mādhava Nidāna* deals with diagnostics whereas *Śārngadhara Samhitā* deals with Pharmacy. Similarly, *Bhāvaprakāśa* is the Materia-medica of Ayurveda.

Problems in the Current System of Ayurveda Education

Today, India officially recognizes Ayurveda and other systems of indigenous medicine along with the conventional biomedicine. To patronize and promote these systems, the Government of India, in 1995, established a separate department for Indian Systems of Medicine and Homeopathy (ISM&H), which is now known as AYUSH (Ayurveda, Yoga, Unani, Siddha, Homeopathy). Among all the systems in AYUSH, presently Ayurveda holds a prominent position and a major share in the infrastructural facilities in terms of the number of hospitals, dispensaries, educational institutions and registered medical practitioners. The Central Council of Indian Medicine (CCIM), which was established through Indian Medicine Central Council Act of 1970, is the governing body that monitors the matters related to Ayurveda education in India. At present, more than 240 Ayurvedic colleges offer a graduate level degree: 'Ayurvedacharya' [Bachelor of Ayurvedic Medicine and Surgery (BAMS)] in India. This course is of 5.5 years' duration after grade 12 with science subjects. Even though the CCIM has imposed various educational norms and regulations, the standard of Ayurvedic education has been a cause of great concern in recent years.

Almost no methodical and authentic work based on data has been published in recent years on the problems related to Ayurvedic education. Therefore, authentic papers/ reports are not available on the topic. However, a few scholars have written some articles which are published in some souvenirs of conferences/magazines and books. Also, some articles are posted on a few websites. Most of these are the reflections of individual perceptions and therefore tend to have biased opinions. However, the summarised views of these scholars have been presented in the following paragraphs under various problem-areas.

Mushroom growth of poorly equipped colleges and Privatisation:

The mushrooming of Ayurvedic colleges as a result of the liberal policies of the State Governments and the loopholes in the existing acts are the most important factors that are being held responsible for the erosion in the standards of education. Privatization of the education system is yet another trend that accompanied this phenomenon of mushrooming. For example, at present, the total number of private colleges offering graduate level education in Ayurveda is 186; whereas the number of Government colleges offering the same level of education is only 54. However, most of the undergraduate and postgraduate colleges in the Government sector also suffer from a variety of infrastructural constraints. The following figure shows the growth in the number of Ayurvedic colleges in India

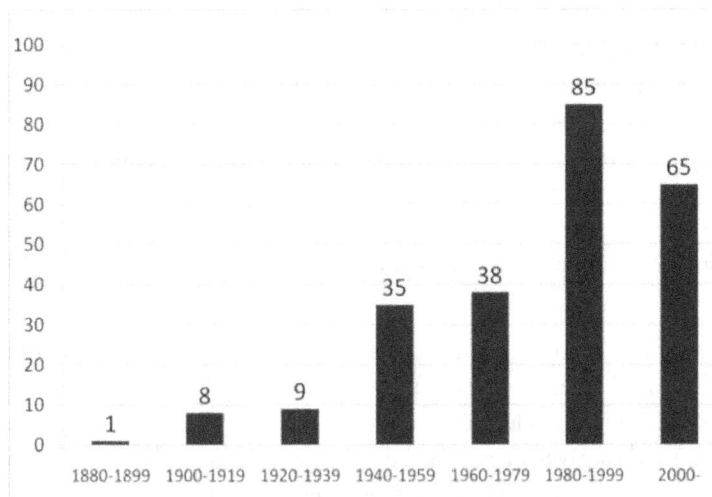

during last 130 years. This diagram is based on the number of the existing colleges and their years of establishment. It is evident from the figure that the majority of these colleges were established during the last 30 years. In many of these colleges, good infrastructure in terms of equipped lecture theatres, laboratories, libraries, operation theatres, Panchakarma facilities and adequate number of qualified teaching and non-teaching staff is not available. This has lead to a dilution in the quality of clinical training.

Joshi VK opines that the mushroom-growth of Ayurvedic Colleges without proper facilities has proven detrimental to the field of Ayurveda. He says that such institutions are basically run by money – minded managements. Students join such institutions after paying a heavy capitation fee. Graduates of such institutes are not useful for the society because they do not have a sound knowledge in the science. Many private and Govt. institutions do not have basic facilities as per the Indian Medicine Central Council Act, 1970 to the extent that some of the institutions are running in single rooms. Only few teachers are working in such institutions who teach multiple subjects. Also, many private organizations have started short term courses in Ayurveda without having any background of Ayurvedic education not only in India, but also in other parts of the world (Joshi VK, 2003).

Dwivedi M and Gupta SN are of view that good infrastructure in terms of spacious and equipped class rooms, laboratories, well-equipped departments, hospital with well equipped labour room, operation theatre and *Pancakarma* facilities are not available in many of the Ayurvedic education institutions (Dwivedi M and Gupta SN, 2008).

According to Sharma BN, most of the Ayurvedic colleges lack the necessary infrastructure, buildings and laboratories etc. Institutional libraries are deficient in optimum number of books and magazines. Laboratories are not given importance in many institutions. Most of the colleges do not have gardens enriched with different medicinal plants (Sharma BN, 2008).

Gehlot S and Singh BM observe that mushrooming of private Ayurvedic Colleges is taking place without having the basic infrastructure, teaching, paramedical and ministerial staff. Therefore in this scenario, for the maintenance of optimum standards of education, there should be sufficient teaching, non teaching staff, equipments and other basics facilities. They also observe that in most of the colleges, laboratory facilities, facilities for drug identification, pharmacy for the drug preparation & their standardization, well established hospital with round the clock emergency facilities are not available (Gehlot S and Singh BM, 2008).

Problems in Understanding Basic Concepts of Ayurveda:

Srinivasulu M states that the fundamental aspects required to understand the patho-physiology in Ayurveda are based on *Doṣa- Dhātu -Mala* theory and imparting of this theory is improper and insufficient in most of the Ayurvedic colleges in the country. Teachers are not able to provide sufficient knowledge to the students regarding the analysis of human physiology and pathology based on *Tridoṣa* theory. According to him, lack of enlightenment of basic subject like *Padārtha Vijanāna*, which is similar in importance to Biochemistry of the modern medical science, is another problem. Also, *Sahitās* are being misinterpreted either due to superstition or with imperfect modern knowledge by teachers (Srinivasulu M, 2006).

Srinivasulu M, further states that one is not able to understand the assessment of the action of a drug in the light of *Rasa- Guna - Vīrya - Vipāka* theory, which is the backbone of Ayurvedic practice. The dilemma of the student or practitioner is alike in the selection of a drug: whether to follow the traditional practices or modern researchers. Inability in diagnosing a case based on Ayurvedic literature and inability to take help from modern laboratorial, technological facilities is another clinical problem. Furthermore, one is unable to use and interpret the modern gadgets into *Doṣa – Dhātu –Mala* theory because only observations of results are taken into consideration rather than extending the understanding into the changes occurring in the body (Srinivasulu M, 2006).

In the opinion of Sridhar N, the documented knowledge in Ayurveda Saṁhitās is only indicatory but not explanatory. Therefore, each statement of Saṁhitā demands furthermore explanation to understand its content. He opines that the various publications available today do not fulfil this need. The books available today are either Saṁhitās as such or a compilation written as per the CCIM syllabus prescribed for undergraduates (Sridhar N, 2006).

Lack of Exposure to Basic Clinical Skills:

According to Sridhar N, Ayurvedic community is not getting adequate clinical exposure. In fact, even the clinical teaching faculty members too are not provided with enough provisions to have adequate clinical exposure because of which many sections of knowledge in Ayurveda are left untouched. Even those sections in which some work has been done are also not developed to the adequate level. As a result, the important basic concepts of Ayurveda and management methods of various diseases have not been analyzed properly and laid down in a uniform manner. It has become very difficult for an Ayurvedic doctor to be confident of assessing prognosis of a pathological condition in a given patient. In turn, it has also become very difficult for teaching faculty to instil confidence among Ayurvedic students regarding the science (Sridhar N, 2006).

Srinivasulu M, while giving his opinion on the issue, states that the students don't witness effective Ayurvedic methods of treatments in the institutions and therefore are not confident in handling the cases when it comes to practice. They are usually unable to give fast relief in common ailments like cough, fever etc. in comparison to allopathic medicines. Also, Ayurvedic physicians are unable to cope up with critical situations of a disease due to lack of knowledge related to prognosis. Further, Srinivasulu M observes that Ayurvedic practitioners are not interested generally in improving their skills by reading journals, attending seminars and by visiting institutions because of a pessimistic attitude that 'there is nothing to improve' (Srinivasulu M, 2006).

According to Sridhar N, there are certain important unique concepts of Ayurveda like *Doṣa-sancāra, Marma, Vyādhighatakas, Rogamārga* and *Kriyākāla*. These concepts are so important that without understanding them one cannot really understand even the sketch of Ayurveda. But unfortunately these concepts are not well analyzed and explained to the student community. Also, these concepts are not put in practice by most of the Ayurvedic professionals. Owing to this it has become very difficult and scaring for an Ayurvedic doctor to understand and manage a complex and critical pathological situation in a given patient (Sridhar N, 2006).

According to Gehlot S and Singh BM, at present, in most of the institutions, students are not exposed to attempt the emergencies in practice. Most of the ayurvedic graduates are not able to interpret the investigation reports of ECG, EEG, Spirometry, USG, CT Scan etc in practice. Therefore, the proper training of emergency handling, clinical pathology must be a part of curriculum (Gehlot S and Singh BM).

As per the observations of Patwardhan K and Gehlot S, in the existing system of undergraduate training, students are generally not exposed to a large variety of cases as patients visiting Ayurvedic institutions belong to only few identifiable categories. Also, students are generally not trained in clinical skills and procedures like incision and drainage, suturing, catheterization, intra venous infusion etc incapacitating them from becoming confidant clinicians who can offer primary healthcare. Students are generally not trained to handle the routine clinical emergencies through Ayurvedic methods. This lack of sufficient clinical exposure forces them to learn some modern medicine anywhere outside the institution and indulge in the practice of Allopathy (Patwardhan K and Gehlot S, 2008).

Problems related to the availability of Ayurvedic Literature:

Sridhar N opines that except for the two sets of classical books, i.e., – *Bṛhat Trayī & Laghu Trayī,* so many other important textbooks of Ayurveda are not easily available to the academic zones. Such textbooks include - *Cakradatta, Sahasrayoga, Vṛndamādhava, Gadanigraha, Vaṅgasena, Cikitsākalikā,* and *Kakṣaputatantra.* The knowledge contained in these books has been neglected since many years and needs to be utilised (Sridhar N, 2006).

Perception of Ayurveda as Philosophical and Impractical System:

Sridhar N also states that owing to the conversational method of narration and certain story-like content of Saṁhitās, Ayurveda is considered as more historical and philosophical than practical and scientific. This misconception has generated reluctant attitude towards the science and has failed to evoke analytical thinking among students and practitioners. For example, the concepts of *Marma,* concept of *Ojas* and concept of *Mānasa Roga* (especially *Graha Rogas*) have been neglected without understanding their actual implications (Sridhar N, 2006).

Poor Dissemination of Recent Advances:

According to Sridhar N, the content of the Ayurvedic science understood so far is relatively less. Even this amount of information is not uniformly available among all Ayurvedic professionals themselves. Because of this, though a lot of work is being done and many young scientists are coming up with appreciable innovative knowledge on various aspects of the science, their findings have been confined to a limited regional belt of professionals and their true validation has remained doubtful. This is happening because there is no one to take care of the knowledge and see that it spreads in a uniform manner among all Ayurvedic professionals across the country. This situation is ultimately rendering the entire work go in vain and is hampering the interest of scholars (Sridhar N, 2006).

According to Patwardhan K and Gehlot S, the essential basic information of recent studies/ reports related to efficacy of Ayurvedic medicines/ procedures is not included in the curriculum, which is required to be included. Also, certain essential topics like Clinical decision making, Care of hospitalized patients, Patient interviewing skills, Ethical decision making and Geriatric patient care are required to be emphasized in the curricula of the clinical disciplines. Certain essential topics related to medical practice including Cost effective medical practice, Quality assurance in medicine, Practice management and Medical record-keeping are to be included in the curriculum (Patwardhan K and Gehlot S, 2008).

Reluctant Attitude of Governing Bodies:

Sridhar N observes that various institutional, Government and autonomous bodies like CCIM are not giving adequate importance and paying attention towards the system. It is only in the very recent past, according to him, that the Government has taken up certain steps in the interest of the system by increasing funds and bringing Ayurvedic dispensaries under one shelter together with allopathic hospitals (Sridhar N, 2006).

Dwivedi M and Gupta SN state that regulatory bodies have failed in ensuring the quality education in the Ayurvedic colleges. According to them, the approval of new colleges is not matching the requirements of the country. In few states like Maharashtra and Karnataka there is mushrooming of the colleges while in the states like Bihar and Assam only few colleges are there (Dwivedi M and Gupta SN, 2008).

Poor Quality of Teachers:

Dwivedi M and Gupta SN opine that the selection criteria that are followed while recruiting teachers usually do not assess one's liking towards teaching profession and his/her ability to teach. Rather, they are based mainly on formal educational qualifications and the degree a person possesses (Dwivedi M and Gupta SN, 2008).

At present, there is no NET/GATE like eligibility test to assure the entry of good scholars into the teaching profession and just anyone who has completed his/her postgraduate level of education is eligible to be a teacher. This situation is discouraging because the aptitude and communication skills that are essential for a good teacher are not evaluated at all. Especially in private colleges, the standard procedures like floating an open advertisement and selection of the teachers by a panel of experts are not followed. These institutes do not pay their teachers the standard salaries as recommended by the UGC. No regulatory authority is there to monitor the pay packages of teachers. In some states, the situation is even worse. Mere submission of one's certificates at the institution is sufficient to get the recognition as a teacher at that institution. Such teachers show their presence in the college only on the occasions of inspection by the CCIM and are never involved in the process of actual teaching and training. To curb this practice, the inspecting authorities must be very strict while recognising such institutions. Also, a uniform pay package to all teachers of Ayurvedic colleges is to be made mandatory by the Government. Promotion opportunities for all teachers are also required to be monitored (Patwardhan K and Gehlot S, 2008).

Poorly Structured Curriculum:

Dwivedi M and Gupta SN state that the contents of Ayurvedic curriculum are not clear with respect to legal implications of certain practices/ procedures included in the syllabus. Also, they opine that un-thoughtful integration (mere translation of words in to Sanskrita) has proved detrimental to the knowledge and practice of Ayurveda. One of the major compulsions for changing the syllabus is with the intention to make it more useful for the community. But policy-making bodies till date have not able to decide the limitations of various subjects, especially when the matter is related to the integration of allopathy. The most undesirable consequences of this, however, would be production of Ayurvedic graduates who will be lacking confidence in either system of medicine. According to Dwivedi M and Gupta SN, some system-neutral subjects like anatomy, physiology, radiology, hospital management should be accepted as such without making any demarcation label as Ayurvedic or modern (Dwivedi M and Gupta SN, 2008).

Gehlot S and Singh BM opine that there are many controversial topics included in the curriculum. For the betterment of Ayurveda, they suggest *Sandhāya Saṁbhāṣā* to be organized on all the controversial subjects and to arrive at final unanimous conclusions (Gehlot S and Singh BM, 2008).

According to Bhardwaj PK and Upadhyay SD, through extensive discussions controversies over the anatomical organs should be affirmed e.g. *Kloma, Śirohṛdaya* and *Urohṛdaya* and similar other topics. These are some aspects of Ayurvda that require vivid consideration to expand the horizon of Ayurveda and medical education as a whole. There are so many other points which require consideration in the light of modern science to enrich the knowledge of Ayurveda (Bhardwaj PK and Upadhyay SD, 2008).

According to the observations by Patwardhan K and Gehlot S, students generally feel that certain topics in *'Rachana Sharira'* like *'Marma'*, *'Sira'*, *'Snayu'*, *'Sandhi'* etc. are almost outdated as more advanced knowledge on these topics is available in the textbooks of Modern Anatomy/ Modern surgery. But, basic understanding related to some of these structures is very essential to understand other clinical subjects. Therefore, only clinically applicable topics are to be retained in the subject and details related to controversial structures like *'Kloma'* are to be left for post graduate students (Patwardhan K and Gehlot S, 2008).

Similarly, in the subject *'Kriya Sharira'*, topics like 'Assessment of *Prakriti* and *Dhatu Sara'* are given too much of importance whereas the clinical applicability of these topics is not emphasized in clinical disciplines. Therefore, students feel that these topics are only of theoretical importance and not practical. Ideally, teachers of clinical disciplines are to be trained to use these basics in diagnosis and management (Patwardhan K and Gehlot S, 2008).

According to Patwardhan K and Gehlot S, students perceive most of the Ayurvedic topics in *'Agada Tantra'* related to classifications/ numbers/ varieties of poisons and their effects to be outdated and impractical. Such topics are needed to be removed from the curriculum. Also, practical training related to the basics of medical jurisprudence, toxicology and forensic medicine is generally not emphasized in undergraduate teaching making a B.A.M.S. graduate inefficient in handling the legal procedures (Patwardhan K and Gehlot S, 2008).

Patwardhan K and Gehlot S opine that topics related to *'Ariṣṭa Vijñāna'* explained in *'Indriya Sthāna'* of *'Caraka Saṁhitā'* are practically not useful because they do not fit in to the present social scenario. Therefore, the content related to classical *Saṁhitās* also requires to be curtailed based on the current needs and applicability (Patwardhan K and Gehlot S, 2008).

Lacking the advantages of *Gurukula* System of Study:

Muralidhara N opines that there should be certain criteria for selection of students and Guru as explained in Ayurvedic textbooks. According to him, the present day training of the students is entirely different from that of the Gurukula system. Today, complete theoretical aspects are taught in non-clinical subjects in the beginning and then only students are exposed to pre-clinical and finally to the clinical subjects. When one analyzes the classical texts, one can understand that students were exposed to more practical classes with limited and only essential theoretical explanations in ancient medical education system. For example, theoretical aspect of anatomy of *Guda* is described in classical text Suśruta Saṁhitā at the end of *Arśo-nidana* and anatomical relationship of *Hṛdaya* is referred in the context of *Marma Śārīra*. Muralidhara N further opines that, in the process of learning, *Adhyayana, Adhyāpana* and *Tadvidyā Saṁbhāṣā* are considered important methods to be adopted. All these are possible by adopting Gurukula system of education wherein, a disciple comes in direct contact with a Guru. By adopting present day advanced technology there is a big gap created between the student and teacher relationship. The important useful information and techniques which used to be passed on to the student after confirming his/her dedication towards the subject is not seen in present-day education. In understanding such points, only Gurukula system of education can help. (Muralidhara N, 2008).

Poor Teaching Methodology:

As per the observations of Patwardhan K and Gehlot S, the students with science background feel that the current Ayurvedic teaching methodology does not keep up the scientific values and scientific spirit of a young student. This is mainly because the current teaching methodology does not encourage questioning among students. Whatever that is explained in the classical textbooks is perceived to be final and unquestionable. Also, interpretation of theories like *'Tridosha'*, *'Pancha Mahabhuta'*, *'Rasa'*, *'Guṇa'* etc varies largely from one teacher to another making them further confusing and vague. Furthermore, memorizing the reference number of a particular chapter/ verse of any *'Samhita'* does not serve any practical purpose for an undergraduate student, but is given undue importance in teaching and examination system. Memorizing the numbers of various structures / their measurements / various classifications of diseases *(Sankhya Samprapti)* as per different authors etc. also is a futile exercise; which is given undue importance in teaching and examination system (Patwardhan K and Gehlot S, 2008).

According to Gehlot S and Singh BM, the use of audio-visual techniques is to be encouraged for better understanding and presentation of the subject. Teaching should be interactive and questioning between the teacher and students must be encouraged. Clinical teaching/demonstration should be encouraged on the patients (Gehlot S and Singh BM, 2008).

Zollman C and Vickers A (1999) while commenting on Complementary and Alternative systems of medicine, state that Complementary practitioners are not generally concerned with understanding the basic scientific mechanism of their particular therapy. Their knowledge-base is often derived from a tradition of clinical observation and treatment decisions are usually empirical. Sometimes traditional teachings are handed down in a way that discourages questioning and evolution of practice, or that encourages reliance on their own and others' individual anecdotal clinical and intuitive experiences.

Redundant Examination System:

Gehlot S and Singh BM opine that the entrance examination pattern at post graduate level should be applied and problem based. Also, they state that the pattern of examination system at undergraduate level should be changed. In place of descriptive or essay type questions, short answers or multiple choice questions must be included for the assessment of knowledge (Gehlot S and Singh BM, 2008).

Patwardhan K and Gehlot S opine that the examination system in Ayurveda does not assess the actual abilities and skills of a student; rather, it largely depends on the assessment of memorizing capacity of students (Patwardhan K and Gehlot S, 2008). This fact becomes clear when one goes through a few sets of question papers of various universities. Most of these questions are memory-based and do not require any amount of understanding and interpretation. They are mostly incapable of assessing one's knowledge-based analytical skills and problem solving abilities. This type of examination system discourages the innovative thinking process and analytical ability of a student.

Poor training in the issues related to Safety, Efficacy and Quality in Ayurvedic Medicines:

Patwardhan K and Gehlot S observe that in the subject *'Dravyaguna'*, essential basic information related to recent advances in pharmacodynamic/ pharmacognostic/ phytochemical attributes of various Ayurvedic herbs is not emphasized. Basic knowledge related to various technologically advanced methods of 'Drug Standardization' is also not included in the curricula of either *'Dravyaguna'* or *'Rasa Shastra'*. Topics like pharmaco-vigilance, safety profile, toxicity studies and Good Manufacturing Practices are to be ideally included in *'Rasashastra'*. Also, students are not exposed to basic knowledge related to the methods of quantitative and qualitative analysis of chemical components of Ayurvedic preparations. If students are trained in these technical fields, job opportunities for B.A.M.S. degree holders may significantly increase in pharmaceutical industry (Patwardhan K and Gehlot S, 2008).

In the opinion of Gehlot S and Singh BM, identification of the drug is a very difficult job. For the proper identification, fundamentals of Pharmacognosy, phyto-chemistry, methods of evaluation of drug's effects and drug standardization should be a compulsory part of study and research. For this purpose, the subject specialists should be appointed in Ayurvedic institutions (Gehlot S and Singh BM, 2008).

WHO states that many Traditional Medicine (TM) practices and products have been used for a considerable period of time and are therefore considered safe. While this could in general be true, there are a growing number of reports documenting adverse effects that at times have caused deaths. The possible reasons for these adverse effects are many: misuse of the therapies, therapies for which information on safety is lacking, and interaction between TM products and allopathic medicines are some of them. Hence there is a growing need for health care providers to appreciate potential interactions of allopathic and traditional medicines and for consumers to be made aware too (WHO Strategy for Traditional Medicine 2002-2005).

Many TM products and practices have been used for a considerable period of time and are therefore thought to be safe and effective. However, with health care systems increasingly demanding evidence of measurable improvements in health before supporting a product or practice, TM too has to be assessed in a similar manner. Promoting research into the safety and efficacy of TM products and practices is the best means of generating this evidence. (WHO Strategy for Traditional Medicine 2002-2005).

Poor Job Opportunities:

As per the observations of Patwardhan K and Gehlot S, in most of the states, a B.A.M.S. degree holder cannot practice Allopathy legally and therefore hospitals generally prefer M.B.B.S. graduates as medical officers instead of B.A.M.S. graduates. In Ayurvedic educational institutions, only Post Graduate doctors are employed and not B.A.M.S. degree holders. Furthermore, most of the research institutions prefer Post Graduate doctors and therefore, job opportunities in research institutions are limited. Even in Government sector, job opportunities are limited for a B.A.M.S. graduate in certain areas e.g., Railways and Defence. Also, Ayurvedic pharmaceutical firms prefer Post Graduate candidates or pharmaceutical chemists to B.A.M.S. degree holders as experts. Apart from these facts, there is a lot of competition for jobs among B.A.M.S. degree holders as a result of mushrooming of Ayurvedic colleges (Patwardhan K and Gehlot S, 2008).

Other Global Challenges:

WHO in its policy on Traditional Medicines for 2002 – 2005, has stated that there is a need to assure the quality of both the products and the services in Traditional Medicine. A registration system similar to that of western drugs is feasible to ensure that commercially available TM products have correct labelling, contain the correct ingredients and do not overstate efficacy. Manufacturers should be provided guidance and there should be regulations to ensure that appropriate standards are maintained. Commercial manufacture of herbal medicines has some similarity to the manufacture of western drugs; a TM policy could well borrow from the experiences in western medicine (WHO Strategy for Traditional Medicine 2002-2005).

Furthermore, the said WHO document says that Traditional Medicine arising from the experiences of the past and embedded in the culture of each society cannot stand still and must change and develop. Along with allopathic medicine it shares issues in appropriate and rational use. This includes qualification and licensing of providers, proper use of good quality products, good communication between TM providers and patients and provision of scientific information and guidance to the public. The patient is the ultimate beneficiary of any system of medicine and therefore should have access to good scientific information. The provision of such information is a shared responsibility of TM providers, their professional associations and the Government (WHO Strategy for Traditional Medicine 2002-2005).

Regarding the education and training of TM providers, WHO opines that its first priority would be to ensure appropriate training and examination followed by a licensing system. This would be underpinned by legislation thus ensuring that only those that are qualified can practise TM and sell TM products too. If practitioners of other systems are to provide TM services, then their training too should have a component of TM (WHO Strategy for Traditional Medicine 2002-2005).

According to Patwardhan K and Gehlot S, many countries legally do not allow Ayurveda practice and therefore, there are no opportunities for B.A.M.S. graduates in such countries. Only a few Ayurvedic academicians figure in authoring the scientific and evidence based papers in reputed international journals. Ayurvedic academicians generally do not follow international standards while planning the protocols of research projects and while writing research reports. Ayurvedic scholars generally do not have knowledge regarding 'Intellectual Property Rights' and patenting procedures. Authentic websites providing up-to-date knowledge in Ayurveda are not hosted by Ayurvedic institutions. No standard international indexed and peer-reviewed journals are published by Ayurvedic institutions making it difficult for Ayurvedic researches have global attention. Pharmacodynamic/ pharmacokinetic properties/ efficacy/ safety profiles and chemical compositions of Ayurvedic formulations are yet to be established making it difficult for experts in conventional medicine to accept Ayurveda (Patwardhan K and Gehlot S, 2008).

Poor Training related to Entrepreneurship:

Patwardhan K and Gehlot S observe that students are not trained in management skills required to launch a new Ayurvedic hospital/ Panchakarma center/ Ayurvedic Pharmacy during B.A.M.S. course. Students are not exposed to the basics of economical aspects related to healthcare sector during B.A.M.S. course. Students are not introduced to the skills related to the management of Health tourism and emerging opportunities in this field. Students are not exposed to the agricultural and marketing aspects of medicinal plants making it difficult to go for cultivation / marketing of medicinal plants. Also, students are not exposed to the manufacturing techniques related to cosmetic products and such other popular dosage forms during BAMS course making them unfit for modern pharmaceutical industry (Patwardhan K and Gehlot S, 2008).

Poor Quality of Research Publications:

Kessler C points out the following weak points related to research publications in Ayurveda:

- Many publications are only retrievable via hand-search of references and interviews of experts.
- Only a small number of Ayurvedic studies are listed in common western Databases and CAM databases.
- Various studies are published in regional languages, many of them only as abstracts. A large number is not available at all.
- The majority of the studies belong to evidence levels 2nd to 4th. Only very few studies from evidence level Ia to Ib. There are no networks of competence or centers for excellence.
- No publication on health services research (HSR) and health technology assessment (HTA) are available.
- Group sizes are small in Ayurvedic studies. This makes the studies vulnerable for methodological error.
- Rationale for selected study designs is not always properly described. Missing Values complicate the calculation of probability and power (Kessler C, 2006).

Choudhary RR states that the traditional scientific approach of modern clinical trials may not be applicable to Traditional Systems of Medicine (TSM) as it is, due to various inherent problems. WHO has considered this problem and opined that Conventional concepts of clinical research design may be difficult to apply for TSMs. Ayurveda and other TSMs involve and are based on the concerned

Philosophies and fundamental principles; hence any research to be called Ayurvedic research should consider and be based on these fundamental principles. Hence, the choice of study design should be discussed on a case-by-case basis with experienced traditional medical practitioners (Choudhary RR, 2005).

Gehlot S and Singh BM opine that at present no serious efforts are being made in the field of research in most of the institutions. Some research work is continued in few postgraduate institutions but students and even some teachers are not aware about research plan and methodology. Therefore, basic knowledge about the research methodology should be incorporated at undergraduate level which will be helpful in facilitating better research understanding at post graduate level (Gehlot S and Singh BM, 2008).

Baghel MS, while addressing the issue records his observations as follows:

"Some times researcher is hesitant in parting with the data. Practitioners for maintaining their confidentiality do not want to divulge their data. Lack of good research practices, lack of proper clinical research protocol etc. leaves it difficult to collect credible data which is a prime requisite for publication in peer- reviewed journals. Quality of data asked by international journals and reputed Medical journals often cannot be met by Ayurvedic researchers. To document and publish research findings in standard Journals, presentation of material in English is required. In most of the states of India, Ayurvedic teaching is conducted in regional languages hence Ayurvedic research scholars lack command in English language which adds to their problem in preparing research protocols and research papers. Also, there are no incentives and compulsion for publishing the research work in Ayurvedic institutions" (Baghel MS, 2006).

Gehlot S and Singh BM opine that drug research should not be totally based on modern parameters while other fundamental principles of Ayurveda must also be considered. In their opinion, there is great demand for the collaboration between the traditional knowledge base and scientific innovators and researchers. Routine modern methods of drug research are not sufficient for Ayurvedic drugs. For this appropriate methodology has to be developed in consideration of Ayurvedic concepts. The educational institutions, national laboratories and industries have to join hand to come forward with full enthusiasm to launch research on leads provided by traditional practitioners (Gehlot S and Singh BM, 2008).

Zollman C and Vickers A (1999) have enumerated the following factors that limit research in complementary medicine:

Lack of funding—Many funding bodies have been reluctant to give grants for research in complementary medicine. Pharmaceutical companies have little commercial interest in researching complementary medicine.

Lack of research skills—Complementary practitioners usually have no training in critical evaluation of existing research or practical research skills.

Lack of an academic infrastructure—This means limited access to computer and library facilities, statistical support, academic supervision and university research grants.

Insufficient patient numbers—Individual list sizes are small, and most practitioners have no disease "specialty" and therefore see very small numbers of patients with the same clinical condition. Recruiting patients into studies is difficult in private practice.

Difficulty undertaking and interpreting systematic reviews—Many poor quality studies make interpretation of results difficult. Many publications in complementary medicine are not on standard databases such as Medline. Many different types of treatment exist within each complementary discipline.

Methodological issues—Responses to treatment are unpredictable and individual, and treatment is usually not standardised. Designing appropriate controls for some complementary therapies (such as acupuncture, manipulation) is difficult, as is blinding patients to treatment allocation. Allowing for the role of the therapeutic relationship also creates problems.

Issues Related to Integration:

There have been always issues related to integration and mainstreaming of Ayurveda. Questions like 'How much of modern medicine component should be there in Ayurvedic curriculum?', 'How much of modern medicine can be practiced by an ISM&H practitioner?' and 'Is it possible to handle clinical emergencies of primary healthcare level only with Ayurveda without using Allopathy?' need to be answered in clear terms (Tewari PV, 2008).

In this regard, National health Policy in ISM&H (2002), states that:

(a) Efforts would be made to integrate and mainstream ISM&H in health care delivery systems including National Programmes.

(b) A range of options for utilization of ISM&H manpower in the health care delivery system would be developed by assigning specific goal oriented role and responsibility to the ISM work-force. An ISM&H wing would be encouraged and supported at the primary health care level.

(c) States would be encouraged to re-enact or modify laws governing the practice of modern medicine by ISM practitioners so that there is clarity of the subject.

But exact mechanisms of materialising these goals have not yet been devised.

In view of Bodeker G (2001), integration works best when based on self regulation in relation to standards of practice and training. This needs to be matched by a central or regional system for drug control and evaluation and maintenance of good manufacturing practice; this system should also generate and support a comprehensive programme of research. When conventional medicine dominates complementary medicine, loss of essential features of complementary medicine can occur, and professional conflicts can arise. Policy should aim to keep fees for complementary medicine affordable and within reach of all levels of society. Major sectoral investment is a prerequisite for the development of effective services for complementary medicine; underinvestment risks perpetuating poor standards of practice, services and products.

The Study Design

The present study was planned to evaluate the *'Relevance of current system of Ayurvedic education in the emerging global scenario'* with special reference to graduate level Ayurvedic education in India leading to a BAMS (Ayurvedacharya) degree.

Method adopted in the study: Mailed Survey

Population: The population for the present study was defined in terms of students and teachers studying and teaching respectively in Ayurvedic educational institutions of India during the period of September 2005 to October 2008. Specifically the population was as follows:

1. Students:

- All interns / house surgeons registered under BAMS course who have passed the third professional BAMS examinations successfully.

- All Post-Graduate students registered under Ayurveda Vachaspati –MD(Ay) or Ayurveda Dhanvantari -MS(Ay) courses in Ayurvedic educational institutions recognised by CCIM.

2. Teachers:

- All teachers of Ayurvedic colleges/ universities who possess at least BAMS or equivalent degree.

Sample Frame:

The sample frame which became available to the investigator constituted a list of 242 Ayurvedic colleges spread all over 28 states and 7 union territories of India. As the students and teachers in these colleges constituted the primary units of sampling, for random sampling it was essential to get a list of all the students and teachers. But as no such database is available in India, the investigator was compelled to accept the list of 242 Ayurvedic colleges as the sample frame for this study.

Sample: With the availability of the sample frame described above, Random Cluster Sampling technique was considered as most appropriate. Hence, it was planned to include at least 10% of institutions from each zone while trying to include as many states as possible. The colleges from each zone were selected randomly and all the teachers working in them and the students studying in these colleges (as per operational definition described above) were taken as clusters to construct the sample for collecting data for this study. Thus, a total of 32 colleges were included in the study.

Total Number of institutions and states covered in the study from each zone is given below:

	East	North	South	West	Total
Number of Institutions covered	5	10	8	9	32
Number of States covered	4	7	3	4	18

Tool used in the study: A validated semi-structured questionnaire having 8 sections and 73 items constituted the main survey instrument for this study. The details pertaining to its development are as follows:

Methods adopted for Preparing the Questionnaire:

An exhaustive list of items was prepared on the basis of interactions the investigator had with students and teachers of various educational institutions. Various sources of literature- like journals, reports of various committees, National Health Policy, souvenirs of conferences, news reports and other articles were also consulted for collecting the items. These items covered various problem-areas of graduate level Ayurvedic education. These randomly listed items were then re-arranged carefully to form clusters/ sections representing specific problem-areas. Finally, eight such clusters/sections were constructed, under which a total of 73 items were grouped to form a semi-structured questionnaire:

1. Problems related to the exposure of a BAMS graduate to basic clinical skills
2. Problems related to job opportunities after the completion of BAMS course
3. Problems related to the relevance of the Curriculum of BAMS course
4. Problems related to Teaching methodology in the existing system of Ayurvedic education
5. Problems related to Global Challenges being faced by the Ayurvedic syestem of medicine
6. Problems related to Entrepreneurship /Business opportunities after the completion of BAMS course
7. Perception regarding Ideal system of Medical Education for India.
8. Problems related to Personal relevance of Ayurveda to the teachers and students.

Each of the above sections contained multiple statements in simple English language. While framing the questions, care was taken to avoid the influence of biases related to the investigator. Thus, the questions framed were not necessarily representing the views of the investigator, but contained the ideas gathered after thorough literature search and interactions the investigator had with students and teachers of various educational institutions.

All these statements expressed the problems related to respective areas with a negative connotation. The respondents were given the option of recording their response in the form of **'Strongly Agree'**, **'Agree'**, **'Undecided'**, **'Disagree'** and **'Strongly Disagree'** by recording a check mark (√) in the respective columns provided for the purpose, except for Section No.7. Section No.7 contained three statements in the form of 'Multiple Choices' out of which the respondents were asked to choose only one option that they considered was the best.

Apart from the statements related to problem areas, the respondents were also requested to record some basic data related to them. This data included the details such as Name, Age, Sex, Institutional affiliation, Type of Institution (Governmental/Semi-Governmental or Private), State where the institution is situated and the present status of the respondent (Student or Teacher). Students were asked to record the course in which they are enrolled and the teachers were asked to record their academic qualifications, subject speciality and present designation.

Also, the questionnaire contained a declaration statement by the investigator stating the purpose of study and assuring the strict confidentiality related to the identity of respondents. The respondents were also asked to sign on another declaration statement stating that their participation in the study was purely voluntary and the responses given by them were based on their own individual perceptions and that they were not compelled to respond in any particular way by the investigator or by any other authority.

Process of Validation of the Questionnaire:

Initially, the preliminary questionnaire was distributed to 150 respondents in the Faculty of Ayurveda, Banaras Hindu University on a random basis. Respondents included Interns, Post graduate students and teachers. They were given 1 to 2 days of time to fill the questionnaire and return it. The validation process of the questionnaire was carried out on the basis of first 100 completed questionnaires received by the investigator.

The process of testing the questionnaire for its reliability and validity was completed by following the steps stated in the following paragraphs:

1. Data Entry:

The data was fed on to the computer using the software 'Statistical Package for Social Sciences (SPSS)' (Version 11.5). The basic data were fed in the 'String' format and the responses to specific items were fed in 'Numerical' format. For this conversion of responses into numerical format, the following scoring was used for all sections, except for Section No.7:

Strongly Agree	=	5
Agree	=	4
Undecided	=	3
Disagree	=	2
Strongly Disagree	=	1

For section No.7, the options given were serially numbered 1, 2 & 3 and accordingly, the respondents' choice was fed on to the computer on the basis of the serial number of the option.

2. Test for Reliability / Consistency:

Reliability test in the form of Cronbach's coefficient Alpha was carried out for all sections of the questionnaire separately except for section No.7. This was done to find out the Correlation between the respective item and the total sum score (without the respective item) and the internal consistency of the scale (coefficient *alpha*) if the respective item would be deleted. Here, each section was considered as an independent scale. No statistical reliability test could be performed on section No.7 because the section contained only one item which was of 'Multiple choice' type. While validating the scale, value of alpha greater than 0.7 was considered acceptable (J Martin Bland and Douglas G Altman, 1997) and item-total correlation greater than 0.2 was considered acceptable (Streiner D, Norman G, 1995).

Finalising the Questionnaire:

Depending on the feedback received, some minor corrections were made in the pattern/language of the statements. Wherever the value of 'a (when item deleted)' was greater than Cronbach's coefficient a, the corresponding item was deleted and the whole process of validation was repeated and thereafter, the questionnaire was finalized. The finalised full questionnaire is given below:

Data Collection Procedure:

The final validated and tested questionnaire for its reliability, was printed on both sides of A-4 size paper so that each questionnaire contained 3 sheets (6 Pages) and was mailed to about 32 Ayurvedic colleges spread all over India so that at least 10% of all institutes existing in each geographical zone were covered. Also, the aim was to include as many states as possible. Varying number of copies of the questionnaire were mailed to each institution based on several factors like: total admission capacity, presence or absence of post graduate courses, availability of interns/ house surgeons in the institution during the period of study, total number of teachers in the institution etc. Heads of these institutions were requested through a formal letter to distribute the questionnaires among all interns/ house surgeons, Post Graduate students and teachers. Self-addressed stamped envelopes were also mailed for returning the filled questionnaires.

A period of 1 to 2 days was given to the respondents to return the filled questionnaires. The questionnaires thus filled were collected and the data was fed on to the computer using SPSS software. After the completion of the study, the results were analysed using the same software package.

Inclusion Criteria:

Institutions:

- Governmental/ Semi-Governmental and Private Ayurvedic educational institutions spread across India, recognised by Central Council of Indian Medicine and offering at least BAMS course were included in the study. Institutions were selected on the basis of random cluster sampling method as described earlier.

Students:

- Interns / House surgeons enrolled in the selected institutions who have passed the third professional BAMS examinations successfully, irrespective of their age, sex and nationality were included in the study.

- Ayurvedic Post-Graduate students enrolled in the selected institutions irrespective of their age, sex, nationality and subject specialization were included in the study.

Teachers:

- Teachers working in the selected Ayurvedic colleges/ institutions who possess at least BAMS or equivalent degree, irrespective of their age, sex, subject specialization, designation and experience were included in the study.

Exclusion Criteria:

- Students who are registered under BAMS course in the selected institutions but who have not yet passed their third professional BAMS examinations were not included in the study to avoid biased and immature perceptions.
- Teachers working in the selected Ayurvedic educational institutions but not having a BAMS or equivalent degree were excluded from the study (For example, teachers teaching Sanskrit).

The following table shows the 18 states covered in the study after grouping them under respective geographical zones (This classification is based on the Department of tourism, Government of India).

States covered in the study as per the geographical zones

East	North	South	West
Assam	Haryana	Andhra Pradesh	Goa
Bihar	Himachal Pradesh	Kerala	Gujarat
Orissa	Punjab	Karnataka	Maharashtra
West Bengal	Rajasthan		Madhya Pradesh
	Uttarakhand		
	Delhi		
	Uttar Pradesh		

The Response Rate

The response rate for students in this study was 59.6% and for teachers it was 54%. For a study of this nature, an overall response rate of 57.4% can be considered good.

Sample Details:

The following tables depict the characters of the sample including institutional affiliation, age-group, zone-wise and state-wise distribution, male- female participation, status-wise distribution etc.

State-wise Distribution of Participants as per their status						
State	BAMS	PG	Lecturer	Reader	Professor	Total
Andhra Pradesh	17	0	18	0	2	37
Assam	9	20	7	1	1	38
Bihar	0	5	10	3	1	19
Delhi	13	0	6	1	2	22
Goa	31	0	11	1	4	47
Gujarat	2	18	4	0	3	27
Haryana	0	0	16	5	2	23
Himachal Pradesh	9	20	0	0	1	30
Kerala	28	7	17	4	3	59
Karnataka	56	37	28	2	11	134
Maharashtra	0	58	6	3	5	72
Madhya Pradesh	18	23	22	5	2	70
Orissa	12	16	2	5	3	38
Punjab	12	0	9	3	3	27
Rajasthan	7	73	14	0	1	95
Uttarakhand	7	12	22	4	8	53
Uttar Pradesh	36	73	47	21	18	195
West Bengal	15	10	8	2	1	36
Total	272	372	247	60	71	1022

From the figures given in the table above, it is clear that the maximum numbers of participants were from the state of Uttar Pradesh (195) and the minimum were from the state of Bihar (19). Among participants from different categories, maximum numbers of participants were PG students (372) and minimum were Readers (60).

Zone-wise Distribution of Participants as per their status:						
Zone	BAMS	PG	Lecturers	Readers	Professors	Total
East	36	51	27	11	6	**131**
North	84	178	114	34	35	**445**
South	101	44	63	6	16	**230**
West	51	99	43	9	14	**216**
Total	**272**	**372**	**247**	**60**	**71**	**1022**

As the above table suggests, among BAMS students, maximum number of participants were from South zone (101) and among PG students, the maximum participants were from North zone (178). Minimum participants among BAMS students and PG students were from East (36) and South (44) zones respectively. Among Lecturers, maximum numbers of participants were from North zone (114) and minimum were from East zone (27). Among Readers, maximum numbers of participants were from North zone (34) and minimum were from South zone (6). Among Professors, maximum numbers of participants were from North zone (35) and minimum were from East zone (6).

The following table shows the distribution of the institutions covered in the study in different geographical zones on the basis of their type of administration.

Distribution of institutions on the basis of their type of administration				
Zone	Government	Private	Semi-Government	Total
East	5	0	0	5
North	9	1	0	10
South	4	3	1	8
West	5	2	2	9
Total	**23**	**6**	**3**	**32**

The following table shows the Zone-wise Distribution of participants as per their gender and status:

Zone-wise Distribution of participants as per their gender and status							
Zone	Gender	BAMS	PG	Lecturers	Readers	Professors	Total
East	Female	14	23	8	2	0	**47**
	Male	22	28	19	9	6	**84**
North	Female	30	64	26	5	6	**131**
	Male	54	114	88	29	29	**314**
South	Female	64	17	19	3	2	**105**
	Male	37	27	44	3	14	**125**
West	Female	30	43	10	2	3	**88**
	Male	21	56	33	7	11	**128**
Total		**272**	**372**	**247**	**60**	**71**	**1022**

The table above depicts the distribution of participants as per their gender and status. It may be observed that no female professors from east zone participated in the study. Among BAMS students, female participants outnumbered male participants in west and south zones.

Distribution of participants as per their age-group							
Status	Age Group (Years)						Total
	21-30	31-40	41-50	51-60	61-70	71-80	
BAMS	267	4	1	0	0	0	272
PG	324	44	4	0	0	0	372
Lecturers	42	132	54	17	2	0	247
Readers	0	15	27	18	0	0	60
Professors	0	9	15	40	6	1	71
Total	**633**	**204**	**101**	**75**	**8**	**1**	**1022**

A maximum number of participants in the present study belonged to the age group of 21-30 years (633).

Zone-wise Distribution of Participants based on their status and type of institutional affiliation							
ZONE	Affiliation	BAMS	PG	Lecturer	Reader	Professor	Total
East	Govt.	36	51	27	11	6	131
	Total	36	51	27	11	6	**131**
North	Govt.	72	178	105	31	31	417
	Private	12	0	9	3	4	28
	Total	84	178	114	34	35	**445**
South	Govt.	56	7	37	2	7	109
	Private	38	29	23	4	3	97
	Semi Govt.	7	8	3	0	6	24
	Total	101	44	63	6	16	**230**
West	Govt.	20	63	30	5	6	124
	Private	31	8	12	2	5	58
	Semi Govt.	0	28	1	2	3	34
	Total	**51**	**99**	**43**	**9**	**14**	**216**

As the above table is indicating, the participants of Semi-Governmental institutional affiliation were present in the West and South zones only. Similarly, no private colleges were covered from East zone.

Analysis of Data:

The data obtained was fed on to the computer using the software 'Statistical Package for Social Sciences (SPSS)' (Version 11.5). The basic data related to the respondent were fed in the 'String' format and their responses to individual items were fed in 'Numerical' format. For the conversion of responses into numerical format, the following scoring pattern was used for all sections, except for Section No.7:

Strongly Agree	=	**5**
Agree	=	**4**
Undecided	=	**3**
Disagree	=	**2**
Strongly Disagree	=	**1**

For section No.7, the options given were serially numbered 1, 2 & 3 and accordingly, the respondents' choice was fed on to the computer on the basis of the serial number of the option.

Observations and Result

This chapter highlights the important observations of the present study. Observations are presented in the order of sections present in the questionnaire.

Observations Related to Section-1 of the Questionnaire:

Section-1 of the questionnaire was related to the problem area **"Extent of exposure to the basic clinical skills during BAMS Course"**. The participants were asked to go through the list of eleven statements and to indicate their level of agreement with each statement in terms of 'Strongly Agree' (SA), 'Agree' (A), 'Undecided' (U), 'Disagree' (D) and 'Strongly Disagree' (SD). Participants were divided into two groups – 'Student' and 'Teacher'. The 'Student' group included BAMS students (Interns/ House surgeons) and PG Students. 'Teacher' group included Lecturers, Readers and Professors. All the responses obtained were calculated for 'Teacher' and 'Student' groups separately. Following scoring pattern was used to calculate the mean scores:

Strongly Agree (SA)	=	5
Agree (A)	=	4
Undecided (U)	=	3
Disagree (D)	=	2
Strongly Disagree (SD)	=	1

Independent Sample-T test was applied to compare the mean scores of 'Student' (S) and 'Teacher' (T) groups. For interpretation of data, following meanings are drawn from the responses:

- 'Strongly Agree' and 'Agree': **Zone of Agreement**
- 'Undecided': **Neutral Zone**
- 'Disagree' and 'Strongly Disagree': **Zone of Disagreement**

The same methods as explained above were also followed for drawing inferences on other sections of the questionnaire, except for section-7. Following tables summarise the responses of both the groups along with mean scores and results of Independent Samples-T Test for each item. Section-1 contained a total of 11 items. Items are serially numbered from Q1.1 to Q1.11.

The abbreviations used in the tables are as follows: 'Strongly Disagree' **(SD)**, 'Disagree' **(D)**, 'Agree' **(A)**, 'Undecided' **(U)** and 'Strongly Agree' **(SA)**.

Q1.1

Students are not trained to handle the clinical emergencies of primary healthcare level through Ayurvedic methods.

Table-Q1.1:

		SD	D	U	A	SA	Mean ± SD	t	p
Student	N	6	15	4	149	470	4.65 ± 0.702	5.734	**0.000**
(N=644)	%	0.9%	2.3%	0.6%	23.1%	73.0%			
Teacher	N	11	22	5	127	213	4.35 ± 0.977		
(N=378)	%	2.9%	5.8%	1.3%	33.6%	56.3%			
Total	N	17	37	9	276	683			
	%	1.7%	3.6%	0.9%	27.0%	66.8%			

As **Table-Q1.1** suggests 73% of all students and 56.3% of all teachers strongly agree with the statement. Also, of all participants, 93.8% are in the zone of agreement. The percentage of teachers and students who disagree with the statement is very negligible. This means that a significant number of students and teachers tend to agree that students are not trained to handle the clinical emergencies of primary healthcare level through Ayurvedic methods during BAMS course. Also, there is a statistically highly significant difference between the mean scores of student and teacher groups (p<0.001); the mean score being significantly higher in student group. This means that students tend to agree with the statement more strongly than teachers.

Q1.2

Students are not exposed to any successful Ayurvedic method of primary healthcare in the management of infectious conditions like malaria and tuberculosis.

Table Q1.2:

		SD	D	U	A	SA	Mean ± SD	t	p
Student (N=644)	N	10	55	18	217	344	4.29 ± 0.981	3.636	**0.000**
	%	1.6%	8.5%	2.8%	33.7%	53.4%			
Teacher (N=378)	N	12	48	13	143	162	4.04 ± 1.122		
	%	3.2%	12.7%	3.4%	37.8%	42.9%			
Total	N	22	103	31	360	506			
	%	2.2%	10.1%	3.0%	35.2%	49.5%			

As the **Table Q1.2** indicates, 87.1% of students are in the zone of agreement. Number of teachers in Agreement zone is 80.7%. This means that a significant number of students and teachers tend to agree that students are not exposed to any successful Ayurvedic method of primary healthcare level in the management of infectious conditions like malaria and tuberculosis during BAMS course. Also, there is a statistically highly significant difference between the mean scores of student and teacher groups (p<0.001); the mean score being significantly higher in student group. This means that students tend to agree with the statement more strongly than teachers.

Q1.3

Students are not exposed to any successful Ayurvedic method of primary healthcare in the management of poisoning.

Table Q1.3

		SD	D	U	A	SA	Mean ± SD	t	p
Student (N=644)	N	5	14	19	146	460	4.62 ± 0.721	3.228	**0.001**
	%	0.8%	2.2%	3.0%	22.7%	71.4%			
Teacher (N=378)	N	6	11	7	133	221	4.46 ± 0.808		
	%	1.6%	2.9%	1.9%	35.2%	58.5%			
Total	N	11	25	26	279	681			
	%	1.1%	2.4%	2.5%	27.3%	66.6%			

As the **Table Q1.3** suggests, 94.1% of students and 93.7% of teachers are in the zone of agreement. Overall, 93.9% of all participants are in the zone of agreement. Therefore, a significant number of students and teachers tend to agree that students are not exposed to any successful Ayurvedic method of primary healthcare level, in the management of poisoning during BAMS course. Also, there is a statistically highly significant difference between the mean scores of student and teacher groups (p<0.005), the mean score being significantly higher in student group. This means that students tend to agree with the statement more strongly than teachers.

Q1.4
Students are not exposed sufficiently to the basic clinical skills and procedures like incision and drainage, suturing and catheterization.

Table Q1.4

		SD	D	U	A	SA	Mean ± SD	t	p
Student (N=644)	N	65	171	14	176	218	3.48 ± 1.438	2.921	0.004
	%	10.1%	26.6%	2.2%	27.3%	33.9%			
Teacher (N=378)	N	47	110	16	125	80	3.21 ± 1.387		
	%	12.4%	29.1%	4.2%	33.1%	21.2%			
Total	N	112	281	30	301	298			
	%	11.0%	27.5%	2.9%	29.5%	29.2%			

Table Q1.4 shows that 61.2% students and 54.3% teachers are in the zone of agreement. This means that a significant number of students and teachers tend to agree that students are not exposed sufficiently to the basic clinical skills and procedures like incision and drainage, suturing and catheterization during BAMS course. Also, there is a statistically highly significant (p<0.005) difference between the mean scores of student and teacher groups; the higher mean scores being in Student group. Therefore, it can be said that students tend to agree more strongly with the statement than teachers.

Q1.5
Students are not trained sufficiently to conduct normal delivery.

Table Q1.5

		SD	D	U	A	SA	Mean ± SD	t	p
Student (N=644)	N	45	103	27	201	268	3.84 ± 1.306	6.159	0.000
	%	7.0%	16.0%	4.2%	31.2%	41.6%			
Teacher (N=378)	N	34	110	25	120	89	3.32 ± 1.347		
	%	9.0%	29.1%	6.6%	31.7%	23.5%			
Total	N	79	213	52	321	357			
	%	7.7%	20.8%	5.1%	31.4%	34.9%			

Table **Q1.5** suggests that 72.8% of students and 55.2% of teachers are in the zone of agreement. It means that a significant number of teachers and students tend to agree that students are not trained sufficiently to conduct normal delivery during BAMS course. Also, mean score is higher in student group. This difference in mean scores is statistically highly significant (p<0.001).

Q1.6
Students are not exposed to a large variety of cases because patients visiting Ayurvedic institutions belong to only few identifiable categories like those complaining of joint pain, ano-rectal diseases, stroke etc.

Table Q1.6

		SD	D	U	A	SA	Mean ± SD	t	p
Student (N=644)	N	36	119	17	272	200	3.75 ± 1.232	6.133	0.000
	%	5.6%	18.5%	2.6%	42.2%	31.1%			
Teacher (N=378)	N	36	120	16	129	77	3.24 ± 1.342		
	%	9.5%	31.7%	4.2%	34.1%	20.4%			
Total	N	72	239	33	401	277			
	%	7.0%	23.4%	3.2%	39.2%	27.1%			

As the **Table Q1.6** suggests, a total of 73.3% students and 54.5% teachers are in the zone of agreement. Therefore, a significant number of students and teachers tend to agree that students are not exposed to a large variety of cases during BAMS course because patients visiting Ayurvedic institutions belong to only few identifiable categories. Also, mean score is higher in student group than that in teacher group. This difference in mean scores is statistically highly significant (p<0.001). So, students agree with the statement more strongly than teachers.

Q1.7
Students are not exposed sufficiently to the basic modern knowledge of the subjects like Physiology, Pathology, Biochemistry, Pharmacology, Medicine, Paediatrics, Obstetrics & Gynaecology, Eye & ENT and Surgery.

Table Q1.7

		SD	D	U	A	SA	Mean ± SD	t	p
Student (N=644)	N	39	137	32	229	207	3.66 ± 1.287	4.047	0.000
	%	6.1%	21.3%	5.0%	35.6%	32.1%			
Teacher (N=378)	N	32	109	15	148	74	3.33 ± 1.304		
	%	8.5%	28.8%	4.0%	39.2%	19.6%			
Total	N	71	246	47	377	281			
	%	6.9%	24.1%	4.6%	36.9%	27.5%			

As **Table Q1.7** depicts, 67.7% of students and 58.8% of teachers are in the zone of agreement. So, a significant number of students and teachers tend to agree that students are not exposed sufficiently to the basic modern knowledge of the subjects like Physiology, Pathology, Biochemistry, Pharmacology, Medicine, Paediatrics, Obstetrics & Gynaecology, Eye & ENT and Surgery during BAMS course. There is a statistically highly significant difference between the mean scores of two groups (p<0.001); mean score being higher for students. This means that students tend to agree with the statement more strongly than the teachers.

Q1.8
Students are not exposed sufficiently to the basic skills of interpreting ECG, X-Ray and such other diagnostic tools and their clinical utility.

Table Q1.8

		SD	D	U	A	SA	Mean ± SD	t	p
Student (N=644)	N	21	111	19	200	293	3.98 ± 1.212	3.622	0.000
	%	3.3%	17.2%	3.0%	31.1%	45.5%			
Teacher (N=378)	N	19	81	9	156	113	3.70 ± 1.242		
	%	5.0%	21.4%	2.4%	41.3%	29.9%			
Total	N	40	192	28	356	406			
	%	3.9%	18.8%	2.7%	34.8%	39.7%			

As the **Table Q1.8** is indicating, 76.6% of students and 71.2% of teachers are in the zone of agreement. Therefore, a significant number of students and teachers tend to agree that students are not exposed sufficiently to the basic skills of interpreting ECG, X-Ray and such other diagnostic tools and their clinical utility during BAMS course. There is a statistically highly significant difference between the mean scores of two groups ($p<0.001$); mean score being higher for students. Therefore, students tend to agree with the statement more strongly than teachers.

Q1.9
Students are not exposed to the basic skills in the areas like Genetic counselling, Human sexuality, End of life care, Geriatrics and Drug and alcohol abuse.

Table Q1.9

		SD	D	U	A	SA	Mean ± SD	t	p
Student (N=644)	N	7	37	35	234	331	4.31 ± 0.894	5.263	0.000
	%	1.1%	5.7%	5.4%	36.3%	51.4%			
Teacher (N=378)	N	13	43	27	151	144	3.98 ± 1.104		
	%	3.4%	11.4%	7.1%	39.9%	38.1%			
Total	N	20	80	62	385	475			
	%	2.0%	7.8%	6.1%	37.7%	46.5%			

Table Q1.9 suggests that 87.7% of students and 78.0% of teachers are in the zone of agreement. This means that a significant number of students and teachers tend to agree that students are not exposed to the basic skills in the areas like Genetic counselling, Human sexuality, End of life care, Geriatrics and Drug and alcohol abuse during BAMS course. Also, mean score is higher in student group than that in teacher group. This difference in mean scores is statistically highly significant ($p<0.001$). Therefore, students tend to agree with the statement more strongly than teachers.

Q1.10
Students are not trained sufficiently in the basic clinical methods related to *Pancakarma*, *Ksāra Karma*, *Ksāra Sūtra* and *Jalaukāvacarana*.

Table Q1.10

		SD	D	U	A	SA	Mean ± SD	t	p
Student (N=644)	N	75	167	38	224	140	3.29 ± 1.364	2.917	**0.004**
	%	11.6%	25.9%	5.9%	34.8%	21.7%			
Teacher (N=378)	N	47	129	27	114	61	3.03 ± 1.338		
	%	12.4%	34.1%	7.1%	30.2%	16.1%			
Total	N	122	296	65	338	201			
	%	11.9%	29.0%	6.4%	33.1%	19.7%			

Table Q1.10 shows that 56.5 % of students and 46.3% of teachers are in the zone of agreement. The number of teachers in the zone of agreement (46.3%) is almost equal to the number of teachers in zone of disagreement (46.5%). This means that a significant number of students agree that students are not trained sufficiently in the basic clinical methods related to *Pancakarma, Ksāra Karma, Ksāra Sūtra* and *Jalaukāvacarana* during BAMS course. Mean score is higher in student group than that in teacher group. This difference in mean scores is statistically highly significant (p<0.005). Therefore, it can be said that students tend to agree with the statement more strongly than teachers.

Q1.11
Students are not exposed sufficiently to the basic methods of physical examination, diagnosis and management of common clinical conditions, making them non-confident clinicians/ practitioners.

TableQ1.11

		SD	D	U	A	SA	Mean ± SD	t	p
Student (N=644)	N	79	225	25	176	139	3.11 ± 1.403	3.175	**0.002**
	%	12.3%	34.9%	3.9%	27.3%	21.6%			
Teacher (N=378)	N	61	150	12	104	51	2.83 ± 1.353		
	%	16.1%	39.7%	3.2%	27.5%	13.5%			
Total	N	140	375	37	280	190			
	%	13.7%	36.7%	3.6%	27.4%	18.6%			

As the **Table Q1.11** suggests, 47.2% of students are in the zone of disagreement and 48.9% of students are in the zone of agreement. However, as the mean scores for student group is more than 3, it can be said that there is a marginal tendency towards agreement among students that they are not exposed sufficiently to the basic methods of physical examination, diagnosis and management of common clinical conditions, making them non-confident clinicians/ practitioners. 55.8% of teachers are in the zone of disagreement and 41.0% of teachers are in the zone of agreement. Mean score is higher in student group than that in teacher group. This difference in mean scores is statistically highly significant (p<0.05).

Inter Zone comparison of means by One-way ANOVA:

The following table, **Table Q1.A** summarises the results of one-way ANOVA test to find out any inter-zone relationship for the Student group with respect to their responses to all items in the section. As the table suggests, statistically significant differences exist between the mean scores for the students of different zones for all items in section-1. Also, Post Hoc test (LSD) reveals the significant pairs in terms of zones. **In the table, 'S' stands for South, 'W' stands for West, 'N' stands for North and 'E' stands for East.**

Table Q1.A- Students

Item	Mean ± SD				Inter zone Comparison One –way ANOVA		
	East (n=87)	South (n=145)	West (n=150)	North (n=262)	F	p	Post Hoc Test (LSD) Significant pairs
1.1	4.64 ± 0.664	4.50 ± 0.747	4.67 ± 0.711	4.72 ± 0.674	3.071	**0.027**	S Vs W, S Vs N
1.2	4.39 ± 0.894	3.86 ± 1.149	4.50 ± 0.825	4.37 ± 0.921	13.487	**0.000**	S Vs E, S Vs W, S Vs N
1.3	4.63 ± 0.552	4.38 ± 0.913	4.75 ± 0.546	4.67 ± 0.711	7.619	**0.000**	S Vs E, S Vs W, S Vs N
1.4	2.84 ± 1.493	3.57 ± 1.348	3.13 ± 1.500	3.85 ± 1.313	15.563	**0.000**	S Vs E, S Vs W, N Vs E, N Vs W
1.5	2.92 ± 1.557	4.09 ± 1.184	3.51 ± 1.370	4.21 ± 1.020	30.018	**0.000**	E Vs S, E Vs N, E Vs W, S Vs W, N Vs W
1.6	3.70 ± 1.163	3.42 ± 1.342	3.88 ± 1.226	3.87 ± 1.165	4.924	**0.002**	S Vs W, S Vs N
1.7	3.43 ± 1.344	3.63 ± 1.364	3.99 ± 1.074	3.58 ± 1.310	4.625	**0.003**	W Vs E, W Vs S, W Vs N
1.8	3.90 ± 1.162	3.44 ± 1.353	4.27 ± 1.042	4.15 ± 1.142	15.166	**0.000**	S Vs E, S Vs N, S Vs W, E Vs S, E Vs W
1.9	4.26 ± 0.799	4.01 ± 1.047	4.41 ± 0.984	4.44 ± 0.729	8.031	**0.000**	S Vs E, S Vs W, S Vs E
1.10	3.72 ± 1.117	2.77 ± 1.273	2.85 ± 1.490	3.68 ± 1.239	24.739	**0.000**	S Vs E, S Vs N, E Vs W, N Vs W
1.11	3.14 ± 1.407	2.57 ± 1.337	3.15 ± 1.441	3.37 ± 1.338	10.718	**0.000**	S Vs N, S Vs E, S Vs W

The following table, **Table Q1.B** summarises the results of one-way ANOVA test to find out any inter-zone relationship for the Teacher group with respect to their responses to all items in the section. As the table suggests, except for the items Q1.5, Q1.7, Q1.9 and Q1.10, no statistically significant differences exist between the mean scores for teachers of different zones for items in section-1. Also, Post Hoc test (LSD) reveals the significant pairs in terms of zones. In the table, S stands for South, W stands for West, N stands for North and E stands for East.

Table Q.1B- Teachers

Item	Mean ± SD				Inter zone Comparison One –way ANOVA		
	East (n=44)	South (n=85)	West (n=66)	North (n=183)	F	p	Post Hoc Test (LSD) Significant Pairs
1.1	4.18 ± 0.971	4.46 ± 0.765	4.14 ± 1.065	4.41 ± 1.022	2.085	0.102	-
1.2	4.05 ± 0.987	4.19 ± 0.945	3.86 ± 1.201	4.04 ± 1.194	1.038	0.376	-
1.3	4.30 ± 0.668	4.56 ± 0.680	4.38 ± 0.718	4.48 ± 0.913	1.352	0.257	-
1.4	2.73 ± 1.318	3.40 ± 1.433	3.18 ± 1.391	3.26 ± 1.365	2.411	0.067	-
1.5	2.75 ± 1.314	3.54 ± 1.385	3.20 ± 1.459	3.39 ± 1.262	3.840	**0.010**	S Vs E, N Vs E
1.6	3.09 ± 1.428	3.06 ± 1.459	3.11 ± 1.383	3.41 ± 1.236	1.907	0.128	-
1.7	2.98 ± 1.210	3.65 ± 1.251	3.44 ± 1.371	3.22 ± 1.299	3.411	**0.018**	E Vs S, N Vs S
1.8	3.39 ± 1.185	3.73 ± 1.276	3.70 ± 1.324	3.75 ± 1.209	1.065	0.364	-
1.9	3.86 ± 1.025	4.31 ± 0.926	3.92 ± 1.027	3.87 ± 1.200	3.304	**0.020**	S Vs E, S Vs W, S Vs N
1.10	3.11 ± 1.205	2.74 ± 1.320	2.74 ± 1.407	3.26 ± 1.315	4.255	**0.006**	S Vs N, N Vs W
1.11	2.66 ± 1.293	2.76 ± 1.306	2.82 ± 1.467	2.90 ± 1.353	0.444	0.722	-

Observations Related to Section-2 of the Questionnaire:

Section-2 of the questionnaire was related to the problem area **"Job opportunities after the completion of BAMS course"**. There were a total of seven items in this section. Items are serially numbered from Q2.1 to Q2.7. Following tables summarise the responses of both the groups. Tables also show the mean scores and results of Independent Samples-T Test for each item in terms of t value and p value.

Q 2.1

Legally, in most of the states, a BAMS degree holder can not practice Allopathy and therefore hospitals generally prefer MBBS graduates as medical officers instead of BAMS graduates.

Table Q2.1

		SD	D	U	A	SA	Mean ± SD	t	p
Student (N=644)	N	11	42	19	188	384	4.39 ± 0.945	6.732	0.000
	%	1.7%	6.5%	3.0%	29.2%	59.6%			
Teacher (N=378)	N	19	40	24	157	138	3.94 ± 1.142		
	%	5.0%	10.6%	6.3%	41.5%	36.5%			
Total	N	30	82	43	345	522			
	%	2.9%	8.0%	4.2%	33.8%	51.1%			

As the **Table Q2.1** suggests, 88.8% of students and 78.0% of teachers are in the zone of agreement. Therefore, a significant number of students and teachers tend to agree that in most of the states, a BAMS degree holder cannot practice Allopathy legally and therefore, hospitals generally prefer MBBS graduates as medical officers instead of BAMS graduates. Furthermore, the mean score of student group is higher than that of teacher group. This difference in mean scoring is statistically highly significant (p<0.001). This observation means that students tend to agree with the statement more strongly than teachers.

Q 2.2
Ayurvedic hospitals are less in number in comparison to Allopathic ones and therefore job opportunities are limited.

Table Q2.2

		SD	D	U	A	SA	Mean ± SD	t	p
Student (N=644)	N	6	39	16	207	376	4.41 ± 0.876	2.415	**0.016**
	%	0.9%	6.1%	2.5%	32.1%	58.4%			
Teacher (N=378)	N	6	22	7	171	172	4.27 ± 0.881		
	%	1.6%	5.8%	1.9%	45.2%	45.5%			
Total	N	12	61	23	378	548			
	%	1.2%	6.0%	2.3%	37.0%	53.6%			

Table Q2.2 suggests that 90.5% of students and 90.7% of teachers are in the zone of agreement. Therefore, a significant number of students and teachers tend to agree that Ayurvedic hospitals are less in number in comparison to Allopathic ones and therefore job opportunities are limited for a BAMS graduate. Furthermore, mean score is higher for student group than that for teacher group. This difference is statistically highly significant (p<0.05). Therefore, it can be said that students tend to agree with the statement more strongly than the teachers.

Q 2.3
In Ayurvedic educational institutions, only Post Graduate doctors are employed and not BAMS degree holders.

Table Q2.3

		SD	D	U	A	SA	Mean ± SD	t	p
Student (N=644)	N	15	68	29	226	306	4.15 ± 1.064	4.717	**0.000**
	%	2.3%	10.6%	4.5%	35.1%	47.5%			
Teacher (N=378)	N	9	74	22	147	126	3.81 ± 1.165		
	%	2.4%	19.6%	5.8%	38.9%	33.3%			
Total	N	24	142	51	373	432			
	%	2.3%	13.9%	5.0%	36.5%	42.3%			

The **Table Q2.3** suggests that 82.6% of students and 72.2% of teachers are in the zone of agreement. It means that a significant number of students and teachers tend to agree with the statement that in Ayurvedic educational institutions, only Post Graduate doctors are employed and not BAMS degree holders. The mean score for students is higher than the mean score for teachers. This difference in the mean scores is statistically highly significant (p< 0.001). Therefore, it can be said that students tend to agree with the statement more strongly than the teachers.

Q 2.4
Most of the research institutions prefer Post Graduate doctors and therefore, job opportunities in research institutions are limited.

Table Q2.4

		SD	D	U	A	SA	Mean ± SD	t	p
Student (N=644)	N	9	28	11	280	316	4.34 ± 0.830	2.866	0.004
	%	1.4%	4.3%	1.7%	43.5%	49.1%			
Teacher (N=378)	N	4	21	14	199	140	4.19 ± 0.831		
	%	1.1%	5.6%	3.7%	52.6%	37.0%			
Total	N	13	49	25	479	456			
	%	1.3%	4.8%	2.4%	46.9%	44.6%			

As the **Table Q2.4** indicates, 92.6% of students and 89.6% of teachers are in the zone of agreement. Therefore, a significant number of students and teachers tend to agree that job opportunities are limited for BAMS graduates in research institutions. The mean score for students is higher than the mean score for teachers. This difference in the mean scores is statistically highly significant ($p < 0.05$).

Q 2.5
Even in Government sector, BAMS graduates are not treated at par with MBBS graduates and therefore, job opportunities are limited in certain areas e.g., Railways.

Table Q2.5

		SD	D	U	A	SA	Mean ± SD	t	p
Student (N=644)	N	3	9	6	147	479	4.69 ± 0.610	3.808	0.000
	%	0.5%	1.4%	0.9%	22.8%	74.4%			
Teacher (N=378)	N	1	12	8	121	236	4.53 ± 0.718		
	%	0.3%	3.2%	2.1%	32.0%	62.4%			
Total	N	4	21	14	268	715			
	%	0.4%	2.1%	1.4%	26.2%	70.0%			

From the **Table Q2.5**, it is clear that 74.4% of students and 62.4% of teachers 'Strongly Agree' with the statement. The mean score for students is significantly higher than the mean score for teachers ($p < 0.001$). Therefore, it may be said that a significant number of students and teachers tend to agree with the statement that in Government sector, BAMS graduates are not treated at par with MBBS graduates and therefore, job opportunities are limited in certain areas e.g., Railways.

Q 2.6
Ayurvedic pharmaceutical firms prefer Post Graduate candidates to BAMS degree holders as experts.

Table Q2.6

		SD	D	U	A	SA	Mean ± SD	t	p
Student (N=644)	N	3	30	35	294	282	4.28 ± 0.803	3.476	0.001
	%	0.5%	4.7%	5.4%	45.7%	43.8%			
Teacher (N=378)	N	3	32	23	191	129	4.09 ± 0.898		
	%	0.8%	8.5%	6.1%	50.5%	34.1%			
Total	N	6	62	58	485	411			
	%	0.6%	6.1%	5.7%	47.5%	40.2%			

As the **Table Q2.6** suggests, 89.5% of students and 84.6% of teachers are in the zone of agreement with the statement that Ayurvedic pharmaceutical firms prefer Post Graduate candidates to BAMS degree holders as experts. The mean score for students is significantly higher than the mean score for teachers ($p< 0.05$).

Q 2.7 There is lot of competition for jobs among BAMS degree holders as a result of mushrooming of Ayurvedic colleges.

Table Q2.7

		SD	D	U	A	SA	Mean ± SD	t	p
Student (N=644)	N	11	43	45	246	299	4.21 ± 0.955	1.140	0.255
	%	1.7%	6.7%	7.0%	38.2%	46.4%			
Teacher (N=378)	N	3	40	32	130	173	4.14 ± 1.010		
	%	0.8%	10.6%	8.5%	34.4%	45.8%			
Total	N	14	83	77	376	472			
	%	1.4%	8.1%	7.5%	36.8%	46.2%			

The **Table Q2.7** suggests that 84.6% of students and 80.2% of teachers are in the zone of agreement. No statistically significant difference exists in between the mean scores of both the groups. Therefore, it may be said that a significant number of teachers and students in the study tend to agree that mushrooming of Ayurvedic colleges has led to considerable competition for jobs among BAMS degree holders.

Inter Zone comparison of means by One-way ANOVA:

The following table, **Table Q2.A** summarises the results of one-way ANOVA test to find out any inter-zone relationship for the Student group with respect to their responses to all items in the section. As the table suggests, statistically significant differences exist between the mean scores for the students of different zones for 5 items in section-2. Also, Post Hoc test (LSD) reveals the significant pairs in terms of zones.

Table Q2.A-Students

	Mean ± SD				Inter zone comparison One-way ANOVA		
Item	East (n=87)	South (n=145)	West (n=150)	North (n=262)	F	p	Post Hoc Test (LSD) Significant Pairs
2.1	4.37 ± 0.864	4.23 ± 1.054	4.45 ± 0.973	4.44 ± 0.885	1.748	0.156	-
2.2	4.54 ± 0.744	4.02 ± 1.070	4.67 ± 0.746	4.43 ± 0.793	15.762	**0.000**	S Vs E, S Vs W, S Vs N
2.3	4.08 ± 1.112	3.86 ± 1.326	4.37 ± 0.973	4.20 ± 0.889	6.227	**0.000**	E Vs W, S Vs W, S Vs N
2.4	4.29 ± 0.926	4.10 ± 1.026	4.41 ± 0.860	4.46 ± 0.604	6.389	**0.000**	S Vs W, S Vs N
2.5	4.63 ± 0.631	4.60 ± 0.776	4.77 ± 0.604	4.72 ± 0.484	2.438	0.064	-
2.6	4.16 ± 0.834	4.08 ± 0.886	4.40 ± 0.769	4.35 ± 0.742	5.461	**0.001**	E Vs W, S Vs W, S Vs N
2.7	3.78 ± 1.176	4.15 ± 1.036	4.38 ± 0.841	4.29 ± 0.843	8.436	**0.000**	E Vs N, E Vs S, E Vs W, S Vs W

The following table, **Table Q2.B** summarises the results of one-way ANOVA test to find out any inter-zone relationship for the Teacher group with respect to their responses to all items in the section. As the table suggests, except for the item 2.2, no statistically significant differences exist between the mean scores for teachers of different zones. Post Hoc test (LSD) reveals the significant pairs in terms of zones for the item 2.2.

Table Q2.B - Teachers

Item	Mean ± SD				Inter zone comparison One-way ANOVA		
	East (n=44)	South (n=85)	West (n=66)	North (n=183)	F	p	Post Hoc Test Significant Pairs
2.1	3.91 ± 1.197	3.81 ± 1.239	3.76 ± 1.359	4.07 ± 0.978	1.743	0.158	-
2.2	4.34 ± 0.805	3.99 ± 1.096	4.29 ± 0.818	4.38 ± 0.782	4.090	**0.007**	S Vs N, S Vs E, S Vs W
2.3	3.93 ± 1.129	3.60 ± 1.347	3.83 ± 1.184	3.87 ± 1.069	1.278	0.282	-
2.4	4.18 ± 0.724	4.08 ± 0.848	4.24 ± 0.703	4.22 ± 0.889	0.665	0.574	-
2.5	4.45 ± 0.791	4.55 ± 0.764	4.45 ± 0.661	4.57 ± 0.699	0.606	0.612	-
2.6	4.00 ± 0.863	4.13 ± 0.842	4.02 ± 0.832	4.11 ± 0.957	0.398	0.755	-
2.7	3.89 ± 1.061	4.07 ± 1.100	4.21 ± 0.903	4.20 ± 0.988	1.405	0.241	-

Observations related to the Section-3 of the questionnaire:

Section-3 of the questionnaire contained 19 items and was related to the problem area **"Scientific relevance of the Curriculum of BAMS course"**. Following tables summarise the responses of both the groups. Tables also show the mean scores and results of Independent Samples-T Test for each item in the section-3 in terms of t value and p value. Items are serially numbered from Q3.1 to Q3.19.

Q3.1

Most of the topics in the subject 'Ayurvedīya Itihāsa' have got least practical applicability.

Table Q3.1

		SD	D	U	A	SA	Mean ± SD	t	p
Student	N	20	53	34	199	338	4.21 ± 1.069	6.838	**0.000**
(N=644)	%	3.1%	8.2%	5.3%	30.9%	52.5%			
Teacher	N	23	60	26	161	108	3.72 ± 1.209		
(N=378)	%	6.1%	15.9%	6.9%	42.6%	28.6%			
Total	N	43	113	60	360	446			
	%	4.2%	11.1%	5.9%	35.2%	43.6%			

As Table **Q3.1** suggests, 83.4% of students and 71.2% of teachers are in the zone of agreement with the statement. Therefore, a significant number of students and teachers tend to agree that most of the topics in the subject 'Ayurvedīya Itihāsa' have got least practical applicability. Furthermore, the mean score is higher for the student group in comparison to teacher group and this difference is statistically highly significant ($p<0.001$).

Q3.2
Most of the topics covered in the subject *'Padārtha Vijnyāna'* are philosophical and their practical applicability is limited.

Table Q3.2

		SD	D	U	A	SA	Mean ± SD	t	p
Student (N=644)	N	46	125	40	230	203	3.65 ± 1.295	4.933	**0.000**
	%	7.1%	19.4%	6.2%	35.7%	31.5%			
Teacher (N=378)	N	44	105	22	134	73	3.23 ± 1.350		
	%	11.6%	27.8%	5.8%	35.4%	19.3%			
Total	N	90	230	62	364	276			
	%	8.8%	22.5%	6.1%	35.6%	27.0%			

Table Q3.2 suggests that more than 65% of students and more than 50% of teachers are in the zone of agreement. Also, there is a statistically highly significant difference between the mean scores of student and teacher groups ($p<0.001$); the mean score being significantly higher in student group. This means that a significant number of students and teachers tend to perceive most of the topics covered in the subject *'Padārtha Vijnyāna'* to be philosophical and to be of limited practical applicability.

Q3.3
Many topics in *'Rachanā Śārīra'* like *'Marma'*, *'Sirā'*, *'Snāyu'*, *'Sandhi'* etc. are outdated as more advanced knowledge on these topics is available in the textbooks of Modern Anatomy/ Modern surgery.

Table Q3.3

		SD	D	U	A	SA	Mean ± SD	t	p
Student (N=644)	N	76	171	56	176	165	3.28 ± 1.399	2.739	**0.006**
	%	11.8%	26.6%	8.7%	27.3%	25.6%			
Teacher (N=378)	N	59	112	27	116	64	3.04 ± 1.381		
	%	15.6%	29.6%	7.1%	30.7%	16.9%			
Total	N	135	283	83	292	229			
	%	13.2%	27.7%	8.1%	28.6%	22.4%			

As **Table Q3.3** indicates, 52.9% of students and 47.6% of teachers are in the zone of agreement. However, mean scores for both the groups are greater than 3 indicating a general tendency towards agreement. Therefore, it can be said that there is a tendency towards agreement that many topics in *'Rachanā Śārīra'* like *'Marma'*, *'Sirā'*, *'Snāyu'*, *'Sandhi'* etc. are outdated as more advanced knowledge on these topics is available in the textbooks of Modern Anatomy/ Modern surgery. The mean score of student group is significantly higher than the mean score of teacher group ($p<0.05$).

Q3.4
Topics like 'Assessment of *Prakrti* and *Dhātu Sāra*' are given undue importance in the subject *'Kriyā Śārīra'* and the clinical applicability of these topics is not emphasized in clinical disciplines.

Table Q3.4

		SD	D	U	A	SA	Mean ± SD	t	p
Student (N=644)	N	71	164	46	231	132	3.29 ± 1.338	2.641	0.008
	%	11.0%	25.5%	7.1%	35.9%	20.5%			
Teacher (N=378)	N	54	116	16	136	56	3.06 ± 1.353		
	%	14.3%	30.7%	4.2%	36.0%	14.8%			
Total	N	125	280	62	367	188			
	%	12.2%	27.4%	6.1%	35.9%	18.4%			

The **Table Q3.4** suggests that just more than 55% of students and just more than 50% of teachers are in the zone of agreement. However, as the mean scores for both the groups are greater than 3, it may be said that there is a tendency towards agreement. The mean score of student group is significantly higher than the mean score of teacher group (p<0.05). Therefore, a significant number of students and teachers in the present study, tend to think that topics like 'Assessment of *Prakrti* and *Dhātu Sāra*' are given undue importance in the subject *'Kriyā Śārīra'* and the clinical applicability of these topics is not emphasized in clinical disciplines.

Q 3.5
The essential practical exposure to the laboratory diagnostic methods in serology, immunology, histopathology, microbiology and parasitology is not emphasized in *'Roga nidāna* and *Vikrti Vijnāna'*.

Table Q3.5

		SD	D	U	A	SA	Mean ± SD	t	p
Student (N=644)	N	8	74	23	271	268	4.11 ± 1.006	3.088	0.002
	%	1.2%	11.5%	3.6%	42.1%	41.6%			
Teacher (N=378)	N	9	53	21	176	119	3.91 ± 1.068		
	%	2.4%	14.0%	5.6%	46.6%	31.5%			
Total	N	17	127	44	447	387			
	%	1.7%	12.4%	4.3%	43.7%	37.9%			

As the **Table Q3.5** indicates, more than 80% of students and more than 75% of teachers are in the zone of agreement with the statement that the essential practical exposure to the laboratory diagnostic methods in serology, immunology, histopathology, microbiology and parasitology is not emphasized in *'Roga nidāna* and *Vikrti Vijnāna'*. The mean scores of student group (4.11 ± 1.006) are significantly higher than the mean scores of teacher group (P<0.05), indicating a stronger tendency towards agreement among students than teachers.

Q 3.6

In *'Dravyaguna'*, essential basic information related to recent advances in pharmacodynamic/ pharmacognostic/ phytochemical attributes of various Ayurvedic herbs and methods of evaluation of their pharmacological effects is not emphasized.

Table Q3.6

		SD	D	U	A	SA	Mean ± SD	t	p
Student (N=644)	N	15	56	33	265	275	4.13 ± 1.010	3.780	**0.000**
	%	2.3%	8.7%	5.1%	41.1%	42.7%			
Teacher (N=378)	N	15	43	26	183	111	3.88 ± 1.078		
	%	4.0%	11.4%	6.9%	48.4%	29.4%			
Total	N	30	99	59	448	386			
	%	2.9%	9.7%	5.8%	43.8%	37.8%			

The **Table Q3.6** shows that more than 80% of students and more than 75% of teachers are in the zone of agreement with reference to the statement. The mean score is significantly higher in student group than in the teacher group ($p > 0.001$). Therefore, a significant number of students and teachers tend to think that essential basic information related to recent advances in pharmacodynamic/ pharmacognostic/ phytochemical attributes of various Ayurvedic herbs and methods of evaluation of their pharmacological effects is not emphasized in *'Dravyaguna'*.

Q3.7

Essential basic knowledge related to various technologically advanced methods of 'Drug Standardization' is not included in the curricula of either *'Dravyaguna'* or *'Rasa Śāstra'*.

Table Q3.7

		SD	D	U	A	SA	Mean ± SD	t	p
Student (N=644)	N	10	44	28	270	292	4.23 ± 0.927	2.898	**0.004**
	%	1.6%	6.8%	4.3%	41.9%	45.3%			
Teacher (N=378)	N	6	46	19	161	146	4.04 ± 1.033		
	%	1.6%	12.2%	5.0%	42.6%	38.6%			
Total	N	16	90	47	431	438			
	%	1.6%	8.8%	4.6%	42.2%	42.9%			

Table Q3.7 suggests that 87.2% of students and 81.2% of teachers are in the zone of agreement with reference to the statement. Mean score of student group is significantly higher than the mean score of the teacher group ($p > 0.05$). Therefore, a significant number of students and teachers in the study tend to agree that essential basic knowledge related to various technologically advanced methods of 'Drug Standardization' is not included in the curricula of either *'Dravyaguṇa'* or *'Rasa Śāstra'*.

Q3.8
Essential basic knowledge related to pharmaco-vigilance, safety profile, toxicity studies and Good Manufacturing Practices– is not included in *'Rasa Śāstra'*.

Table Q3.8

		SD	D	U	A	SA	Mean ± SD	t	p
Student (N=644)	N	6	76	36	261	265	4.09 ± 1.008	0.870	0.384
	%	.9%	11.8%	5.6%	40.5%	41.1%			
Teacher (N=378)	N	9	39	21	170	139	4.03 ± 1.026		
	%	2.4%	10.3%	5.6%	45.0%	36.8%			
Total	N	15	115	57	431	404			
	%	1.5%	11.3%	5.6%	42.2%	39.5%			

Table Q3.8 suggests that more than 80% of students and teachers are in the zone of agreement with the statement. There is no statistically significant difference between the mean scores of the two groups. This means that most of the teachers and students feel that basic knowledge related to pharmaco-vigilance, safety profile, toxicity studies and Good Manufacturing Practices should be included in *'Rasa Śāstra'*.

Q3.9
Essential basic knowledge related to the methods of quantitative and qualitative analysis of chemical components of Ayurvedic preparations is not included in the curriculum of *'Rasa Śāstra'*.

Table Q3.9

		SD	D	U	A	SA	Mean ± SD	t	p
Student (N=644)	N	10	71	45	238	280	4.10 ± 1.038	0.164	0.870
	%	1.6%	11.0%	7.0%	37.0%	43.5%			
Teacher (N=378)	N	4	31	20	196	127	4.09 ± 0.898		
	%	1.1%	8.2%	5.3%	51.9%	33.6%			
Total	N	14	102	65	434	407			
	%	1.4%	10.0%	6.4%	42.5%	39.8%			

Table Q3.9 suggests that 80.5% of students and 85.5% of teachers are in the zone of agreement. No statistically significant difference exists between the mean scores of both the groups. It means that there is a strong tendency towards agreement that essential basic knowledge related to the methods of quantitative and qualitative analysis of chemical components of Ayurvedic preparations is not included in the curriculum of *'Rasa Śāstra'*.

Q3.10

In *'Agada Tantra'*, most of the Ayurvedic topics describing the classifications/numbers/varieties of poisons and their effects are outdated and impractical.

Table Q3.10

		SD	D	U	A	SA	Mean ± SD	t	p
Student (N=644)	N	22	119	49	252	202	3.77 ± 1.176	1.828	0.068
	%	3.4%	18.5%	7.6%	39.1%	31.4%			
Teacher (N=378)	N	14	75	41	156	92	3.63 ± 1.159		
	%	3.7%	19.8%	10.8%	41.3%	24.3%			
Total	N	36	194	90	408	294			
	%	3.5%	19.0%	8.8%	39.9%	28.8%			

As **Table Q3.10** suggests, 70.5% of students and 65.6% of teachers are in the zone of agreement with the statement that most of the Ayurvedic topics describing the classifications/numbers/varieties of poisons and their effects are outdated and impractical. No statistically significant difference exists between the mean scores of both the groups.

Q3.11

Topics related to *'Arishta Vijñāna'* explained in *'Indriya Sthāna'* of *'Caraka Samhitā'* are practically not useful because they do not fit in to the present social scenario.

Table Q3.11

		SD	D	U	A	SA	Mean ± SD	t	p
Student (N=644)	N	32	106	109	187	210	3.68 ± 1.225	7.192	0.000
	%	5.0%	16.5%	16.9%	29.0%	32.6%			
Teacher (N=378)	N	40	113	55	110	60	3.10 ± 1.282		
	%	10.6%	29.9%	14.6%	29.1%	15.9%			
Total	N	72	219	164	297	270			
	%	7.0%	21.4%	16.0%	29.1%	26.4%			

Table Q3.11 suggests that 61.6% of students and 45% of teachers are in the zone of agreement. However, mean scores are greater than 3 for both the groups indicating a tendency towards agreement. Therefore, it can be said that a significant number of teachers and students tend to think that the topics related to *'Arishta Vijñāna'* explained in *'Indriya Sthāna'* of *'Caraka Samhitā'* are practically not useful as they do not fit in to the present social scenario. Mean scores are higher in student group than that in the teacher group. This difference is statistically highly significant ($p<0.001$).

Q3.12

The detailed explanations related to the preparation / measurements of instruments used in *'Panchakarma'* (e.g., *'Basti Netra'*, *'Basti Puṭaka'*) / their defects / complications of wrong use etc. are not practically useful and therefore, are not relevant.

Table Q3.12

		SD	D	U	A	SA	Mean ± SD	t	p
Student (N=644)	N	130	229	52	121	112	2.78 ± 1.414	0.014	0.989
	%	20.2%	35.6%	8.1%	18.8%	17.4%			
Teacher (N=378)	N	56	152	31	99	40	2.78 ± 1.414		
	%	14.8%	40.2%	8.2%	26.2%	10.6%			
Total	N	186	381	83	220	152			
	%	18.2%	37.3%	8.1%	21.5%	14.9%			

As the **Table Q3.12** suggests, the mean scores of both student and teacher groups are less than 3. This means that both groups tend to disagree with the statement that the detailed explanations related to the preparation / measurements of instruments used in *'Panchakarma'* (e.g., *'Basti Netra'*, *'Basti Puṭaka'*) / their defects / complications of wrong use etc. are not practically useful.

Q3.13

Practical training related to the basics of medical jurisprudence, toxicology and forensic medicine is not emphasized in teaching making a BAMS graduate inefficient in handling the legal procedures.

Table Q3.13

		SD	D	U	A	SA	Mean ± SD	t	p
Student (N=644)	N	16	77	29	228	294	4.10 ± 1.093	2.246	**0.025**
	%	2.5%	12.0%	4.5%	35.4%	45.7%			
Teacher (N=378)	N	11	51	15	174	127	3.94 ± 1.085		
	%	2.9%	13.5%	4.0%	46.0%	33.6%			
Total	N	27	128	44	402	421			
	%	2.6%	12.5%	4.3%	39.3%	41.2%			

As the **Table Q3.13** suggests, 81.1% of students and 79.6% of teachers are in the zone of agreement. The mean score is more than 4 in the student group indicating the stronger tendency towards agreement of the group. The mean scores of Teacher group are significantly lower (p<0.05). Therefore, it can be said that a significant number of students and teachers tend to perceive the practical training related to the basics of medical jurisprudence, toxicology and forensic medicine to be deficient and also that this deficiency makes a BAMS graduate inefficient in handling the legal procedures.

Q3.14

Essential information of recent studies/ reports related to efficacy of Ayurvedic medicines/ procedures is not included in the curriculum of clinical disciplines, which is required to be included.

Table Q3.14

		SD	D	U	A	SA	Mean ± SD	t	p
Student (N=644)	N	1	34	24	265	320	4.35 ± 0.799	3.995	0.000
	%	0.2%	5.3%	3.7%	41.1%	49.7%			
Teacher (N=378)	N	2	25	18	206	127	4.14 ± 0.823		
	%	.5%	6.6%	4.8%	54.5%	33.6%			
Total	N	3	59	42	471	447			
	%	0.3%	5.8%	4.1%	46.1%	43.7%			

As the **Table Q3.14** suggests, 90.8% of students and 88.1% of teachers are in the zone of agreement. Therefore, a significant number of students and teachers tend to think that the essential information of recent studies/ reports related to efficacy of Ayurvedic medicines/ procedures is not included in the curriculum of clinical disciplines, which is required to be included. However, the mean score is significantly lower in the teacher group in comparison to student group (p<0.001).

Q3.15

The curricula of clinical disciplines contain many outdated methods of treatment/management which are impractical (e.g., *Dronī Prāveśika Rasāyana*).

Table Q3.15

		SD	D	U	A	SA	Mean ± SD	t	p
Student (N=644)	N	24	85	71	242	222	3.86 ± 1.140	6.179	0.000
	%	3.7%	13.2%	11.0%	37.6%	34.5%			
Teacher (N=378)	N	14	94	51	163	56	3.40 ± 1.123		
	%	3.7%	24.9%	13.5%	43.1%	14.8%			
Total	N	38	179	122	405	278			
	%	3.7%	17.5%	11.9%	39.6%	27.2%			

Table Q3.15 shows that 72.1% students and 57.9% teachers are in the zone of agreement with the statement that the curricula of clinical disciplines contain many outdated methods of treatment/management which are impractical. However, the mean score is significantly lower in the teacher group (p<0.001).

Q3.16
Certain essential topics like Clinical decision making, Care of hospitalized patients, Patient interviewing skills, Ethical decision making and Geriatric patient care are not emphasized in the curricula of the clinical disciplines, which are required to be included.

Table Q3.16

		SD	D	U	A	SA	Mean ± SD	t	p
Student (N=644)	N	16	58	35	291	244	4.07 ± 1.007	2.032	0.042
	%	2.5%	9.0%	5.4%	45.2%	37.9%			
Teacher (N=378)	N	12	33	18	218	97	3.94 ± 0.968		
	%	3.2%	8.7%	4.8%	57.7%	25.7%			
Total	N	28	91	53	509	341			
	%	2.7%	8.9%	5.2%	49.8%	33.4%			

As the **Table Q3.16** suggests, more than 80% of students and teachers are in the zone of agreement with the statement that certain essential topics like Clinical decision making, Care of hospitalized patients, Patient interviewing skills, Ethical decision making and Geriatric patient care are not emphasized in the curricula of the clinical disciplines, which are required to be included. Mean scores are significantly higher in student group in comparison to teacher group ($p<0.05$).

Q3.17
Certain essential topics related to medical practice including Cost effective medical practice, Quality assurance in medicine, Practice management and Medical record-keeping are not included in the curriculum, which are required to be included.

Table Q3.17

		SD	D	U	A	SA	Mean ± SD	t	p
Student (N=644)	N	14	42	32	291	265	4.17 ± 0.944	1.559	0.119
	%	2.2%	6.5%	5.0%	45.2%	41.1%			
Teacher (N=378)	N	5	26	16	220	111	4.07 ± 0.852		
	%	1.3%	6.9%	4.2%	58.2%	29.4%			
Total	N	19	68	48	511	376			
	%	1.9%	6.7%	4.7%	50.0%	36.8%			

Table Q3.17 suggests that 86.3% of students and 87.6% of teachers are in the zone of agreement. Mean scores of both groups are greater than 4 indicating a tendency towards strong agreement. However, no significant difference in mean scores is found in between student and teacher groups.

Q3.18
Many modern technical terms are translated into 'Sanskrit' in the curriculum (e.g., *'Unduka Puccha Śotha'* for Appendicitis, *'Hrtkāryacakra'* for cardiac cycle etc.) which does not serve any practical purpose.

Table Q3.18

		SD	D	U	A	SA	Mean ± SD	t	p
Student (N=644)	N	16	96	59	204	269	3.95 ± 1.153	1.916	0.056
	%	2.5%	14.9%	9.2%	31.7%	41.8%			
Teacher (N=378)	N	18	56	25	160	119	3.81 ± 1.170		
	%	4.8%	14.8%	6.6%	42.3%	31.5%			
Total	N	34	152	84	364	388			
	%	3.3%	14.9%	8.2%	35.6%	38.0%			

As the **Table Q1.18** suggests, more than 70% of students and teachers are in the zone of agreement. There is no significant difference between the mean scores of student and teacher groups. Therefore, students and teachers generally tend to perceive the translation of modern technical terms into Sanskrit to be a waste exercise.

Q3.19
Many controversial topics (e.g., certain structures in *Racanā Śārīra*, certain herbs in *Dravyaguna*) are included in the curricula which lead only to confusion among students.

Table Q3.19

		SD	D	U	A	SA	Mean ± SD	t	p
Student (N=644)	N	14	73	42	246	269	4.06 ± 1.063	3.084	**0.002**
	%	2.2%	11.3%	6.5%	38.2%	41.8%			
Teacher (N=378)	N	14	46	33	176	109	3.85 ± 1.084		
	%	3.7%	12.2%	8.7%	46.6%	28.8%			
Total	N	28	119	75	422	378			
	%	2.7%	11.6%	7.3%	41.3%	37.0%			

As the **Table Q3.19** suggests, 80% of students and 75.4% of teachers are in the zone of agreement. There is a strong tendency of students towards agreement zone as the mean score for student group is greater than 4. Also, the mean score for student group is significantly higher than the mean score of teacher group ($p<0.01$). Therefore, a very significant number of students and teachers tend to agree that the controversial topics (e.g., certain structures in *Racanā Śārīra*, certain herbs in *Dravyaguṇa*) included in the curricula of BAMS lead only to confusion among students.

Inter Zone comparison of means by One-way ANOVA:

The following tables, **Table Q3.A** and table **Q3.B** summarise the results of one-way ANOVA test to find out any inter-zone relationship for the Student group and Teacher group respectively as far as their responses to all items in the Section - 3 are concerned. As the table Q3.A suggests, statistically significant differences exist between the mean scores of students belonging to different zones for 15 items in the section. Also, Post Hoc test (LSD) reveals the significant pairs in terms of zones.

However, as the **Table Q3.B** suggests, except for the item 3.18, no statistically significant differences exist between the mean scores for teachers of different zones for items in section-3. Post Hoc test (LSD) reveals the significant pairs in terms of zones.

Table Q3.A- Students

Q	Mean ± SD				Inter Zone Comparison One-way ANOVA		
	East (n=87)	South (n=145)	West (n=150)	North (n=262)	F	p	Post Hoc Test (LSD) Significant Pairs
3.1	4.23 ± 1.064	4.19 ± 1.074	4.13 ± 1.113	4.27 ± 1.044	0.655	0.580	-
3.2	3.55 ± 1.255	3.24 ± 1.411	3.52 ± 1.365	3.98 ± 1.111	11.893	**0.000**	N Vs E, N Vs W, N Vs S
3.3	3.24 ± 1.446	3.00 ± 1.434	2.90 ± 1.451	3.68 ± 1.231	13.350	**0.000**	N Vs E, N Vs W, N Vs S
3.4	3.38 ± 1.287	3.10 ± 1.342	3.27 ± 1.423	3.38 ± 1.298	1.488	0.217	-
3.5	4.10 ± 1.023	3.87 ± 1.162	4.29 ± 0.900	4.15 ± 0.943	4.547	**0.004**	S Vs W, S Vs N
3.6	4.15 ± 1.040	3.98 ± 1.090	4.16 ± 1.069	4.19 ± 0.912	1.492	0.216	-
3.7	4.15 ± 1.084	4.00 ± 1.014	4.31 ± 0.919	4.33 ± 0.798	4.668	**0.003**	S Vs W, S Vs N
3.8	4.06 ± 1.082	3.77 ± 1.116	4.19 ± 0.967	4.22 ± 0.904	7.026	**0.000**	S Vs N, S Vs E, S Vs W
3.9	4.11 ± 1.061	3.70 ± 1.186	4.24 ± 0.960	4.23 ± 0.928	10.053	**0.000**	S Vs N, S Vs E, S Vs W
3.10	3.66 ± 1.189	3.46 ± 1.202	3.70 ± 1.263	4.01 ± 1.056	7.835	**0.000**	N Vs E, N Vs W, N Vs S
3.11	3.86 ± 1.193	3.43 ± 1.212	3.39 ± 1.299	3.92 ± 1.140	9.135	**0.000**	E Vs W, N Vs W, S Vs E, S Vs N
3.12	2.87 ± 1.379	2.11 ± 1.131	2.63 ± 1.490	3.19 ± 1.374	20.765	**0.000**	S Vs W, S Vs E, S Vs N, N Vs W
3.13	3.84 ± 1.170	3.72 ± 1.165	4.37 ± 1.019	4.24 ± 0.994	12.638	**0.000**	W Vs E, N Vs E, S Vs W, S Vs N
3.14	4.39 ± 0.737	4.12 ± 0.927	4.44 ± 0.764	4.41 ± 0.741	5.137	**0.002**	S Vs N, S Vs E, S Vs W
3.15	3.80 ± 1.160	3.57 ± 1.229	3.72 ± 1.210	4.11 ± 0.984	8.536	**0.000**	N Vs E, N Vs W, N Vs S
3.16	3.95 ± 1.120	3.76 ± 1.049	4.21 ± 1.025	4.20 ± 0.889	7.668	**0.000**	N Vs E, N Vs S, S Vs W
3.17	4.24 ± 0.876	3.70 ± 1.106	4.29 ± 1.012	4.33 ± 0.727	16.261	**0.000**	S Vs E, S Vs W, S Vs N
3.18	3.92 ± 1.174	3.96 ± 1.160	3.90 ± 1.230	3.99 ± 1.100	0.232	0.874	-
3.19	4.03 ± 1.050	3.81 ± 1.247	3.99 ± 1.135	4.25 ± 0.869	5.704	**0.001**	N Vs S, N Vs W

Table Q3.B-Teachers

Q	Mean ± SD				Inter Zone Comparison One-way ANOVA		
	East (n=44)	South (n=85)	West (n=66)	North (n-183)	F	P	Post Hoc Test (LSD) Significant Pairs
3.1	3.66 ± 1.160	4.02 ± 1.012	3.73 ± 1.103	3.58 ± 1.319	2.622	0.050	-
3.2	3.34 ± 1.200	3.16 ± 1.430	3.21 ± 1.376	3.24 ± 1.345	0.172	0.916	-
3.3	3.09 ± 1.411	2.93 ± 1.361	3.11 ± 1.458	3.05 ± 1.364	0.252	0.860	-
3.4	3.00 ± 1.347	3.31 ± 1.319	3.05 ± 1.318	2.97 ± 1.381	1.222	0.301	-
3.5	3.75 ± 1.123	4.07 ± 1.044	3.86 ± 1.094	3.89 ± 1.055	1.044	0.373	-
3.6	3.68 ± 1.137	3.96 ± 0.957	3.85 ± 1.153	3.90 ± 1.092	0.701	0.552	-
3.7	4.02 ± 1.023	4.11 ± 0.964	4.14 ± 0.943	3.99 ± 1.099	0.454	0.714	-
3.8	3.89 ± 1.039	4.12 ± 0.969	3.97 ± 0.976	4.05 ± 1.068	0.602	0.614	-
3.9	3.91 ± 0.960	4.09 ± 0.895	4.11 ± 0.879	4.12 ± .894	0.669	0.572	-
3.10	3.32 ± 1.360	3.62 ± 1.234	3.58 ± 1.138	3.72 ± 1.071	1.494	0.216	-
3.11	3.20 ± 1.322	2.95 ± 1.299	3.27 ± 1.284	3.08 ± 1.264	0.889	0.447	-
3.12	3.02 ± 1.338	2.61 ± 1.319	2.86 ± 1.380	2.76 ± 1.203	1.129	0.337	-
3.13	3.80 ± 1.153	3.71 ± 1.153	3.98 ± 1.116	4.07 ± 1.009	2.464	0.062	-
3.14	4.18 ± 0.620	4.22 ± 0.762	4.20 ± 0.789	4.07 ± 0.902	0.862	0.461	-
3.15	3.20 ± 1.091	3.40 ± 1.177	3.32 ± 1.192	3.49 ± 1.079	0.919	0.431	-
3.16	4.14 ± 0.734	4.04 ± 0.993	3.94 ± 0.926	3.85 ± 1.016	1.445	0.229	-
3.17	4.30 ± 0.668	4.12 ± 0.837	3.97 ± 0.960	4.04 ± 0.854	1.507	0.212	-
3.18	3.50 ± 1.248	4.16 ± 1.045	3.71 ± 1.212	3.75 ± 1.162	4.020	**0.008**	S Vs N, S Vs E, S Vs W
3.19	3.80 ± 1.069	3.95 ± 1.101	3.85 ± 1.056	3.81 ± 1.095	0.378	0.769	-

Observations Related to Section-4 of the Questionnaire:

Section-4 of the questionnaire was related to the problem area **"Teaching methodology that is followed in existing system of Ayurvedic education".** There were a total of twelve items in this section. Items are serially numbered from Q4.1 to Q4.12. Following tables summarise the responses of both the groups.

Q4.1
Ayurvedic teaching methodology does not keep up the scientific values and scientific spirit of a young student.

Table Q4.1

		SD	D	U	A	SA	Mean ± SD	t	p
Students	N	45	125	32	182	260	3.76 ± 1.342	2.987	**0.003**
	%	7.0%	19.4%	5.0%	28.3%	40.4%			
Teachers	N	25	87	26	153	87	3.50 ± 1.254		
	%	6.6%	23.0%	6.9%	40.5%	23.0%			
Total	N	70	212	58	335	347			
	%	6.8%	20.7%	5.7%	32.8%	34.0%			

As the **Table Q4.1** depicts, 68.7% of students and 63.5% of teachers are in the zone of agreement. Therefore, a significant number of students and teachers tend to agree that Ayurvedic teaching methodology does not keep up the scientific values and scientific spirit of a young student. The mean scores are significantly higher in student group in comparison to teacher group (p<0.005).

Q4.2
Ayurvedic teaching methodology does not encourage questioning among the young students.

Table Q4.2

		SD	D	U	A	SA	Mean ± SD	t	p
Students	N	57	159	42	187	199	3.48 ± 1.377	3.337	**0.001**
	%	8.9%	24.7%	6.5%	29.0%	30.9%			
Teachers	N	28	127	25	139	59	3.20 ± 1.261		
	%	7.4%	33.6%	6.6%	36.8%	15.6%			
Total	N	85	286	67	326	258			
	%	8.3%	28.0%	6.6%	31.9%	25.2%			

Table Q4.2 suggests that 59.9% of students and 52.4% of teachers are in the zone of agreement. Therefore, a significant number of students and teachers tend to agree that the teaching methodology followed in Ayurvedic educational institutions does not encourage questioning among the young students. The mean scores for student group are significantly higher than the mean scores of teacher group (p<0.005).

Q4.3
Interpretation of theories like *'Tridosa'* or *'Pancha Mahābhūta'* varies largely from one teacher to another making them further confusing and vague.

Table Q4.3

		SD	D	U	A	SA	Mean ± SD	t	p
Students	N	26	114	45	240	219	3.80 ± 1.201	1.384	0.167
	%	4.0%	17.7%	7.0%	37.3%	34.0%			
Teachers	N	11	72	24	187	84	3.69 ± 1.103		
	%	2.9%	19.0%	6.3%	49.5%	22.2%			
Total	N	37	186	69	427	303			
	%	3.6%	18.2%	6.8%	41.8%	29.6%			

Table Q4.3 suggests that 71.3% of students and 71.7% of teachers are in the zone of agreement. This means that a significant number of teachers and students tend to think that the interpretation of theories like *'Tridosa'* or *'Pancha Mahābhūta'* varies largely from one teacher to another making these theories further confusing and vague.

Q4.4
Memorizing the classical 'Sanskrit' verses is unduly emphasized in Ayurvedic method of teaching, making the process of learning difficult.

Table Q4.4

		SD	D	U	A	SA	Mean ± SD	t	p
Students	N	66	145	24	206	203	3.52 ± 1.396	5.571	0.000
	%	10.2%	22.5%	3.7%	32.0%	31.5%			
Teachers	N	45	131	28	117	57	3.03 ± 1.319		
	%	11.9%	34.7%	7.4%	31.0%	15.1%			
Total	N	111	276	52	323	260			
	%	10.9%	27.0%	5.1%	31.6%	25.4%			

As the **Table Q4.4** indicates, 63.5% of students are in the zone of agreement. Number of teachers agreeing (46.1%) and disagreeing (46.6%) with the statement is almost equal. However, mean scores for both groups are greater than 3. Therefore, it can be said that a significant number of students tend to agree that memorizing the classical 'Sanskrit' verses is unduly emphasized in Ayurvedic method of teaching, making the process of learning difficult. Furthermore, mean scores are significantly higher for student group (p<0.001) in comparison to teacher group.

Q4.5
Memorizing the reference number of a particular chapter/ verse of any 'Samhitā' does not serve any practical purpose, but is given undue importance in teaching and examination system.

Table Q4.5

		SD	D	U	A	SA	Mean ± SD	t	p
Students	N	26	106	37	183	292	3.95 ± 1.238	7.130	**0.000**
	%	4.0%	16.5%	5.7%	28.4%	45.3%			
Teachers	N	33	96	23	152	74	3.37 ± 1.288		
	%	8.7%	25.4%	6.1%	40.2%	19.6%			
Total	N	59	202	60	335	366			
	%	5.8%	19.8%	5.9%	32.8%	35.8%			

As the **Table Q4.5** indicates, 73.7% of students are in the zone of agreement. Number of teachers agreeing with the statement is 59.8%. Therefore, it can be said that a significant number of students and teachers tend to agree that memorizing the reference number of a particular chapter/ verse of any 'Samhitā' does not serve any practical purpose, but is given undue importance in teaching and examination system. Furthermore, mean scores are significantly higher for student group (p<0.001) in comparison to teacher group indicating a stronger tendency of students towards agreement.

Q4.6
The examination system in Ayurveda does not assess the actual abilities and skills of a student; rather, it largely depends on the assessment of memorizing capacity of students.

Table Q4.6

		SD	D	U	A	SA	Mean ± SD	t	p
Students	N	10	50	24	216	344	4.30 ± 0.968	9.179	**0.000**
	%	1.6%	7.8%	3.7%	33.5%	53.4%			
Teachers	N	14	74	24	174	92	3.68 ± 1.150		
	%	3.7%	19.6%	6.3%	46.0%	24.3%			
Total	N	24	124	48	390	436			
	%	2.3%	12.1%	4.7%	38.2%	42.7%			

As the **Table Q4.6** indicates, 86.9% of students are in the zone of agreement. Number of teachers agreeing with the statement is 70.3%. Therefore, it can be said that a significant number of students and teachers tend to agree that the examination system in Ayurveda does not assess the actual abilities and skills of a student; rather, it largely depends on the assessment of memorizing capacity of students. Furthermore, mean scores are significantly higher for student group (p<0.001) in comparison to teacher group indicating a stronger tendency of students towards agreement.

Q4.7
Standard textbooks covering all the topics in curricula are not available in the market, making understanding on the subjects more difficult.

Table Q4.7

		SD	D	U	A	SA	Mean ± SD	t	p
Students	N	12	76	22	247	287	4.12 ± 1.052	5.985	0.000
	%	1.9%	11.8%	3.4%	38.4%	44.6%			
Teachers	N	10	70	25	189	84	3.71 ± 1.088		
	%	2.6%	18.5%	6.6%	50.0%	22.2%			
Total	N	22	146	47	436	371			
	%	2.2%	14.3%	4.6%	42.7%	36.3%			

Table Q4.7 suggests that 83% of students and 72.2% of teachers are in the zone of agreement. Therefore, a significant number of students and teachers tend to agree that standard textbooks covering all the topics in curricula are not available in the market, making understanding on the subjects more difficult. The mean scores for student group are significantly higher than the mean scores of teacher group (p<0.001). Therefore, students tend to agree with the statement more strongly than teachers.

Q4.8
'Sanskrit' is given undue importance in teaching – learning process making the process of understanding the subject difficult.

Table Q4.8

		SD	D	U	A	SA	Mean ± SD	t	p
Students	N	64	199	41	170	170	3.28 ± 1.396	5.916	0.000
	%	9.9%	30.9%	6.4%	26.4%	26.4%			
Teachers	N	64	144	28	102	40	2.76 ± 1.304		
	%	16.9%	38.1%	7.4%	27.0%	10.6%			
Total	N	128	343	69	272	210			
	%	12.5%	33.6%	6.8%	26.6%	20.5%			

As the **Table Q4.8** suggests, teachers tend to disagree with the statement that 'Sanskrit' is given undue importance in teaching – learning process- as the mean score of teacher group for this item is less than 3. A significant number of students (52.8%). are in the zone of agreement. The difference between the two groups in terms of mean scores is statistically highly significant.

Q4.9

Memorizing the numbers of various structures / their measurements / various classifications of diseases *(Sankhyā Samprāpti)* as per different authors etc. does not serve any practical purpose, but is given undue importance in teaching and examination system.

Table Q4.9

		SD	D	U	A	SA	Mean ± SD	t	p
Students	N	24	96	45	244	235	3.89 ± 1.164	5.633	**0.000**
	%	3.7%	14.9%	7.0%	37.9%	36.5%			
Teachers	N	22	97	28	151	80	3.45 ± 1.240		
	%	5.8%	25.7%	7.4%	39.9%	21.2%			
Total	N	46	193	73	395	315			
	%	4.5%	18.9%	7.1%	38.6%	30.8%			

Table Q4.9 suggests that 74.4% of students and 61.1% of teachers are in the zone of agreement. Therefore, a significant number of students and teachers tend to agree that memorizing the numbers of various structures / their measurements / various classifications of diseases *(Sankhyā Sam☐prāpti)* as per different authors etc. does not serve any practical purpose, but is given undue importance in teaching and examination system. The mean scores for student group are significantly higher than the mean scores of teacher group (p<0.001). Therefore, students tend to agree with the statement more strongly than teachers.

Q4.10

Students are not trained sufficiently in certain areas of 'Evidence based medicine' like Interpretation of clinical data and research reports, Literature reviews/critiques, Interpretation of laboratory results and Decision analysis.

Table Q4.10

		SD	D	U	A	SA	Mean ± SD	t	p
Students	N	8	42	31	289	274	4.21 ± 0.896	2.861	**0.004**
	%	1.2%	6.5%	4.8%	44.9%	42.5%			
Teachers	N	12	27	10	214	115	4.04 ± 0.950		
	%	3.2%	7.1%	2.6%	56.6%	30.4%			
Total	N	20	69	41	503	389			
	%	2.0%	6.8%	4.0%	49.2%	38.1%			

Table Q4.10 suggests that 87.4% of students and 87% of teachers are in the zone of agreement. Therefore, a significant number of students and teachers tend to agree that students are not trained sufficiently in certain areas of 'Evidence based medicine' during the BAMS course. The mean scores for student group are significantly higher than the mean scores of teacher group (p<0.005).

Q4.11
Students are not trained in areas like using a computer-based clinical record keeping program, carrying out reasonably sophisticated searches of medical information databases on internet, using a variety of forms of telemedicine etc., making them technologically inferior.

Table Q4.11

		SD	D	U	A	SA	Mean ± SD	t	p
Students	N	15	49	29	215	336	4.25 ± 1.009	1.728	0.084
	%	2.3%	7.6%	4.5%	33.4%	52.2%			
Teachers	N	8	23	17	188	142	4.15 ± 0.914		
	%	2.1%	6.1%	4.5%	49.7%	37.6%			
Total	N	23	72	46	403	478			
	%	2.3%	7.0%	4.5%	39.4%	46.8%			

Table Q4.11 suggests that 85.6% of students and 87.3% of teachers are in the zone of agreement. Therefore, a significant number of students and teachers tend to agree that BAMS students are not trained in areas related to computer-based clinical record keeping program, searching the medical information on internet, using telemedicine etc., making them technologically inferior.

Q4.12
Students are not introduced to essential 'Communication skills' like discussing a prescription error with the patient, providing safe sex counselling, negotiating with a patient who requests unnecessary investigations etc.

Table Q4.12

		SD	D	U	A	SA	Mean ± SD	t	p
Students	N	7	83	49	264	241	4.01 ± 1.032	1.570	0.117
	%	1.1%	12.9%	7.6%	41.0%	37.4%			
Teachers	N	8	45	17	213	95	3.90 ± 0.978		
	%	2.1%	11.9%	4.5%	56.3%	25.1%			
Total	N	15	128	66	477	336			
	%	1.5%	12.5%	6.5%	46.7%	32.9%			

Table Q4.12 suggests that 78.4% of students and 81.4% of teachers are in the zone of agreement. Therefore, a significant number of students and teachers tend to agree that BAMS students are not introduced to essential 'Communication skills' like discussing a prescription error with the patient, providing safe sex counselling, negotiating with a patient who requests unnecessary investigations etc.

Inter Zone comparison of means by One-way ANOVA:

The following table, **Table Q4.A** summarises the results of one-way ANOVA test to find out any inter-zone relationship for the Student group with respect to their responses to all items in the Section-4. As the table suggests, statistically significant differences exist between the mean scores for the students of different zones for all items except for Q4.6 and Q4.7. Also, Post Hoc test (LSD) reveals the significant pairs in terms of zones.

Table Q4.A- Students

Item	Mean ± SD				Inter Zone Comparison One-way ANOVA		
	East (n=87)	South (n=145)	West (n=150)	North (n=262)	f	p	Post Hoc test (LSD) - Significant Pairs
Q4.1	3.97 ± 1.125	3.25 ± 1.382	3.54 ± 1.548	4.09 ± 1.141	15.335	**0.000**	S Vs E, S Vs N, E Vs W, N Vs W
Q4.2	3.62 ± 1.305	2.96 ± 1.374	3.54 ± 1.422	3.70 ± 1.306	9.922	**0.000**	S Vs W, S Vs N, S Vs E
Q4.3	3.91 ± 1.096	3.50 ± 1.370	3.75 ± 1.215	3.95 ± 1.096	4.778	**0.003**	S Vs E, S Vs N
Q4.4	3.71 ± 1.329	3.17 ± 1.458	3.16 ± 1.470	3.86 ± 1.241	12.809	**0.000**	S Vs E, S Vs N, N Vs W, E Vs W
Q4.5	3.99 ± 1.156	3.64 ± 1.352	3.81 ± 1.328	4.18 ± 1.097	6.715	**0.000**	S Vs E, S Vs N, N Vs W
Q4.6	4.26 ± 0.869	4.26 ± 1.048	4.23 ± 1.064	4.36 ± 0.893	0.667	0.573	-
Q4.7	4.10 ± 0.928	4.11 ± 1.179	4.07 ± 1.109	4.16 ± 0.985	0.267	0.849	-
Q4.8	3.41 ± 1.369	2.94 ± 1.386	2.76 ± 1.422	3.73 ± 1.240	21.104	**0.000**	S Vs E, S Vs N, E Vs W, N Vs W
Q4.9	3.91 ± 1.168	3.59 ± 1.250	3.62 ± 1.251	4.19 ± 0.973	12.612	**0.000**	S Vs E, S Vs N, E Vs N, N Vs W
Q4.10	4.24 ± 0.821	4.03 ± 1.017	4.23 ± 0.935	4.29 ± 0.815	2.556	0.054	-
Q4.11	4.41 ± 0.756	4.03 ± 1.204	4.33 ± 0.952	4.28 ± 0.981	3.383	**0.018**	S Vs W, S Vs E, S Vs N
Q4.12	4.10 ± 1.000	3.70 ± 1.082	4.13 ± 1.032	4.08 ± 0.985	5.867	**0.001**	S Vs W, S Vs E, S Vs N

The following table, **Table Q4.B** summarises the results of one-way ANOVA test to find out any inter-zone relationship for the Teacher group with respect to their responses to all items in the section. As the table suggests, except for the items Q4.4, Q4.6, Q4.7 and Q4.12, no statistically significant differences exist between the mean scores for teachers of different zones. Also, Post Hoc test (LSD) reveals the significant pairs in terms of zones.

Table Q4.B-Teachers

Items	Mean ± SD				Inter Zone Comparison One-way ANOVA		
	East (n=44)	South (n=85)	West (n=66)	North (n-183)	f	p	Post Hoc test (LSD) Significant Pairs
4.1	3.25 ± 1.400	3.52 ± 1.221	3.41 ± 1.403	3.59 ± 1.173	1.020	0.384	-
4.2	3.07 ± 1.301	3.13 ± 1.316	3.15 ± 1.339	3.27 ± 1.201	0.484	0.694	-
4.3	3.59 ± 1.106	3.69 ± 1.102	3.88 ± 1.117	3.64 ± 1.099	0.864	0.460	-
4.4	2.75 ± 1.349	2.78 ± 1.294	3.26 ± 1.439	3.13 ± 1.258	2.720	**0.044**	S Vs W, S Vs N, E Vs W
4.5	3.23 ± 1.344	3.21 ± 1.319	3.61 ± 1.311	3.38 ± 1.247	1.354	0.257	-
4.6	3.61 ± 1.224	3.64 ± 1.163	4.05 ± 1.044	3.58 ± 1.145	2.821	**0.039**	S Vs W, N Vs W

4.7	3.91 ± 0.936	3.84 ± 1.045	3.89 ± 1.165	3.53 ± 1.093	3.218	**0.023**	N Vs S, N Vs E, N Vs W
4.8	2.75 ± 1.278	2.52 ± 1.315	2.92 ± 1.428	2.82 ± 1.251	1.462	0.224	-
4.9	3.41 ± 1.282	3.51 ± 1.201	3.33 ± 1.305	3.48 ± 1.231	0.292	0.831	-
4.10	3.95 ± 0.888	4.11 ± 0.887	4.08 ± 1.042	4.02 ± 0.963	0.322	0.810	-
4.11	4.09 ± 0.884	4.27 ± 0.697	4.12 ± 1.000	4.11 ± 0.977	0.692	0.557	-
4.12	3.43 ± 1.043	4.08 ± 0.834	3.97 ± 1.037	3.91 ± 0.974	4.593	**0.004**	E Vs S, E Vs N, E Vs W

Observations Related to Section-5 of the Questionnaire:

Section-5 of the questionnaire was related to the problem area **"Global Challenges being faced by the Ayurvedic syestem of medicine"**. This section contained eleven items in total. Following tables summarise the responses of both the groups.

Q5.1

Serious questions are being raised on the safety profile of Ayurvedic preparations in some countries posing a threat to the Ayurvedic system of Medicine.

Table Q5.1

		SD	D	U	A	SA	Mean ±SD	t	p
Students	N	8	34	28	279	295	4.27 ± 0.867	4.496	**0.000**
	%	1.2%	5.3%	4.3%	43.3%	45.8%			
Teachers	N	10	36	25	179	128	4.00 ± 1.013		
	%	2.6%	9.5%	6.6%	47.4%	33.9%			
Total	N	18	70	53	458	423			
	%	1.8%	6.8%	5.2%	44.8%	41.4%			

Table Q5.1 indicates that more than 85% of students and more than 80% of teachers are in the zone of agreement. Mean scores for both groups are greater than 4 indicating a strong tendency of both the groups towards agreement. However, the mean scores of student group are significantly more than the mean scores of teacher group (p< 0.001), indicating a stronger tendency of student group towards agreement. Therefore, it can be said that a very significant number of teachers and students tend to agree that serious questions are being raised on the safety profile of Ayurvedic preparations in some countries posing a threat to the Ayurvedic system of Medicine.

Q5.2

Standardization of Ayurvedic preparations is still a problem that needs to be addressed.

Table Q5.2

		SD	D	U	A	SA	Mean ±SD	t	p
Students	N	3	15	17	252	357	4.47 ± 0.706	3.338	**0.001**
	%	0.5%	2.3%	2.6%	39.1%	55.4%			
Teachers	N	1	15	11	189	162	4.31 ± 0.738		
	%	0.3%	4.0%	2.9%	50.0%	42.9%			
Total	N	4	30	28	441	519			
	%	0.4%	2.9%	2.7%	43.2%	50.8%			

Table Q5.2 indicates that 94.5% of students and 92.9% of teachers are in the zone of agreement. Mean scores for both groups are greater than 4 indicating a strong tendency of both the groups towards agreement. However, the mean scores of student group are significantly more than the mean scores of teacher group (p<0.005), indicating a stronger tendency of student group towards agreement. Therefore, it can be said that a very significant number of teachers and students perceive the issue of standardization of Ayurvedic preparations to be a problem that needs to be addressed.

Q5.3

In many countries, legally, practicing Ayurveda is not allowed and therefore, there are no opportunities for BAMS graduates in such countries.

Table Q5.3

		SD	D	U	A	SA	Mean ±SD	t	p
Students	N	4	24	50	201	365	4.40 ± 0.833	2.380	**0.018**
	%	0.6%	3.7%	7.8%	31.2%	56.7%			
Teachers	N	0	18	27	168	165	4.27 ± 0.792		
	%	0.0%	4.8%	7.1%	44.4%	43.7%			
Total	N	4	42	77	369	530			
	%	0.4%	4.1%	7.5%	36.1%	51.9%			

Table Q5.2 indicates that 87.9% of students and 88.1% of teachers are in the zone of agreement. Mean scores for both groups are greater than 4 indicating a strong tendency of both the groups towards agreement. However, the mean scores of student group are significantly more than the mean scores of teacher group (p<0.05), indicating a stronger tendency of student group towards agreement. Therefore, it can be said that a very significant number of teachers and students tend to agree that in many countries, practicing Ayurveda is not allowed legally and therefore, there are no opportunities for BAMS graduates in such countries.

Q5.4

Possible entry of foreign universities in India may pose a threat to the existing educational institutions.

Table Q5.4

		SD	D	U	A	SA	Mean ±SD	t	p
Students	N	21	100	97	215	211	3.77 ± 1.158	3.460	**0.001**
	%	3.3%	15.5%	15.1%	33.4%	32.8%			
Teachers	N	17	84	58	129	90	3.51 ± 1.202		
	%	4.5%	22.2%	15.3%	34.1%	23.8%			
Total	N	38	184	155	344	301			
	%	3.7%	18.0%	15.2%	33.7%	29.5%			

As the **Table Q5.4** indicates, 66.2% of students and 57.9% of teachers are in the zone of agreement. The mean scores are significantly higher in student group (p<0.005). Therefore, it can be said that a significant number of students and teachers perceive the possible entry of foreign universities in India to be a threat to the existing educational institutions.

Q5.5
Ayurvedic academicians do not figure anywhere in authoring the scientific and evidence based papers in reputed international journals.

Table Q5.5

		SD	D	U	A	SA	Mean ±SD	t	p
Students	N	13	87	55	256	233	3.95 ± 1.081	1.490	0.136
	%	2.0%	13.5%	8.5%	39.8%	36.2%			
Teachers	N	9	59	24	177	109	3.84 ± 1.081		
	%	2.4%	15.6%	6.3%	46.8%	28.8%			
Total	N	22	146	79	433	342			
	%	2.2%	14.3%	7.7%	42.4%	33.5%			

Table Q5.5 suggests that more than 75% of students and teachers are in the zone of agreement. Furthermore, there is no significant difference in the mean scores for both the groups. Therefore, it can be said that a significant number of students and teachers tend to agree that Ayurvedic academicians do not figure anywhere in authoring the scientific and evidence based papers in reputed international journals.

Q5.6
Ayurvedic academicians do not voluntarily participate in International platforms to present their research data.

Table Q5.6

		SD	D	U	A	SA	Mean ±SD	t	p
Students	N	11	120	69	240	204	3.79 ± 1.131	0.756	0.450
	%	1.7%	18.6%	10.7%	37.3%	31.7%			
Teachers	N	14	67	24	175	98	3.73 ± 1.138		
	%	3.7%	17.7%	6.3%	46.3%	25.9%			
Total	N	25	187	93	415	302			
	%	2.4%	18.3%	9.1%	40.6%	29.5%			

As the **Table Q5.6** suggests, more than 65% of students and teachers are in the zone of agreement. Therefore, it may be said that a significant number of students and teachers tend to agree that Ayurvedic academicians do not voluntarily participate in International platforms to present their research data. Furthermore, there is no significant difference in the mean scores for both the groups.

Q5.7
Ayurvedic academicians do not follow international standards while planning the protocols of research projects and while writing research reports.

Table Q5.7

		SD	D	U	A	SA	Mean ±SD	t	p
Students	N	4	83	130	210	217	3.86 ± 1.045	-0.095	0.924
	%	0.6%	12.9%	20.2%	32.6%	33.7%			
Teachers	N	8	47	36	184	103	3.87 ± 1.020		
	%	2.1%	12.4%	9.5%	48.7%	27.2%			
Total	N	12	130	166	394	320			
	%	1.2%	12.7%	16.2%	38.6%	31.3%			

Table Q5.7 suggests that 66.3% of students and 75.9% of teachers are in the zone of agreement with the statement that Ayurvedic academicians do not follow international standards while planning the protocols of research projects and while writing research reports. Furthermore, there is no significant difference in the mean scores for both the groups.

Q5.8
Ayurvedic scholars generally do not have knowledge regarding 'Intellectual Property Rights' and patenting procedures.

Table Q5.8

		SD	D	U	A	SA	Mean ±SD	t	p
Students	N	6	52	104	276	206	3.97 ± 0.943	0.824	0.410
	%	0.9%	8.1%	16.1%	42.9%	32.0%			
Teachers	N	8	38	34	195	103	3.92 ± 0.975		
	%	2.1%	10.1%	9.0%	51.6%	27.2%			
Total	N	14	90	138	471	309			
	%	1.4%	8.8%	13.5%	46.1%	30.2%			

Table Q5.8 indicates that 74.9% of students and 78.8% of teachers in the study are in the zone of agreement with the statement. Therefore, it can be said that a significant number of students and teachers tend to agree that Ayurvedic scholars generally do not have knowledge regarding 'Intellectual Property Rights' and patenting procedures.

Q5.9
Authentic websites providing up-to-date knowledge in Ayurveda are not hosted by Ayurvedic institutions.

Table Q5.9

		SD	D	U	A	SA	Mean ±SD	t	p
Students	N	9	39	49	266	281	4.20 ± 0.918	0.073	0.942
	%	1.4%	6.1%	7.6%	41.3%	43.6%			
Teachers	N	3	15	19	210	131	4.19 ± 0.769		
	%	0.8%	4.0%	5.0%	55.6%	34.7%			
Total	N	12	54	68	476	412			
	%	1.2%	5.3%	6.7%	46.6%	40.3%			

As the **Table Q5.9** suggests, more than 80% of students and about 90% of teachers are in the zone of agreement. Mean scores of both the groups are above 4 indicating a strong tendency towards agreement. From this observation, it can be said that a very significant number of students and teachers tend to agree that authentic websites providing up-to-date knowledge in Ayurveda are not hosted by Ayurvedic institutions.

Q5.10

No standard international indexed and peer-reviewed journals are published by Ayurvedic institutions making it difficult for Ayurvedic researches have global attention.

Table Q5.10

		SD	D	U	A	SA	Mean ±SD	t	p
Students	N	1	48	53	270	272	4.19 ± 0.885	0.628	0.530
	%	0.2%	7.5%	8.2%	41.9%	42.2%			
Teachers	N	5	22	17	201	133	4.15 ± 0.853		
	%	1.3%	5.8%	4.5%	53.2%	35.2%			
Total	N	6	70	70	471	405			
	%	0.6%	6.8%	6.8%	46.1%	39.6%			

Table Q5.10 suggests that more than 80% of students and more than 85% of teachers are in the zone of agreement with the statement that no standard international indexed and peer-reviewed journals are published by Ayurvedic institutions making it difficult for Ayurvedic researches have global attention. In both the groups, mean scores are greater than 4 indicating a strong tendency towards agreement.

Q5.11

Pharmacodynamic/ pharmacokinetic properties/ efficacy/ safety profiles and chemical compositions of Ayurvedic formulations are yet to be established making it difficult for experts in conventional medicine to accept Ayurveda.

Table Q5.11

		SD	D	U	A	SA	Mean ±SD	t	p
Students	N	9	32	46	258	299	4.25 ± 0.893	1.896	0.058
	%	1.4%	5.0%	7.1%	40.1%	46.4%			
Teachers	N	3	18	28	201	128	4.15 ± 0.809		
	%	0.8%	4.8%	7.4%	53.2%	33.9%			
Total	N	12	50	74	459	427			
	%	1.2%	4.9%	7.2%	44.9%	41.8%			

Table Q5.11 indicates that more than 85% of students and teachers are in the zone of agreement with the statement that Pharmacodynamic/ pharmacokinetic properties/ efficacy/ safety profiles and chemical compositions of Ayurvedic formulations are yet to be established making it difficult for experts in conventional medicine to accept Ayurveda. In both the groups, mean scores are greater than 4 indicating a strong tendency towards agreement.

Inter Zone comparison of means by One-way ANOVA:

The following table, **Table Q5.A** summarises the results of one-way ANOVA test to find out any inter-zone relationship for the Student group with respect to their responses to all items in the section. As the table suggests, statistically significant differences exist between the mean scores for the students of different zones for all items in section-5 except Q5.1 and Q5.2. Also, Post Hoc test (LSD) reveals the significant pairs in terms of zones. In the table, S stands for South, W stands for West, N stands for North and E stands for East.

Table Q5.A-Students

Item	Mean ± SD				Inter Zone Comparison One Way ANOVA		
	East (n=87)	South (n=145)	West (n=150)	North (n=262)	F	p	Post Hoc Test (LSD) Significant Pairs
5.1	4.24 ± 0.821	4.17 ± 0.908	4.19 ± 1.015	4.38 ± 0.753	2.500	0.059	-
5.2	4.43 ± 0.622	4.41 ± 0.795	4.47 ± 0.748	4.51 ± 0.654	0.801	0.493	-
5.3	4.18 ± 0.934	4.30 ± 0.908	4.49 ± 0.784	4.46 ± 0.766	3.754	**0.011**	E Vs W, E Vs N, S Vs W
5.4	3.56 ± 1.168	3.90 ± 1.138	3.68 ± 1.137	3.81 ± 1.171	1.997	**0.113**	E Vs S
5.5	3.79 ± 1.163	3.38 ± 1.208	4.15 ± 0.951	4.19 ± 0.916	22.215	**0.000**	E Vs S, E Vs W, E Vs N, S Vs W, S Vs N
5.6	3.84 ± 0.987	3.34 ± 1.187	3.83 ± 1.149	3.98 ± 1.072	10.663	**0.000**	S Vs E, S Vs W, S Vs N
5.7	3.86 ± 0.930	3.47 ± 1.054	3.81 ± 1.145	4.10 ± 0.947	12.232	**0.000**	S Vs E, S Vs W, S Vs N, N Vs W
5.8	3.83 ± 0.943	3.48 ± 1.042	4.11 ± 0.860	4.21 ± 0.819	22.082	**0.000**	E Vs N, E Vs S, E Vs W, S Vs W, S Vs N
5.9	4.09 ± 0.936	3.94 ± 1.075	4.25 ± 0.899	4.34 ± 0.790	6.821	**0.000**	S Vs W, S Vs N, E Vs N
5.10	4.26 ± 0.842	3.77 ± 1.014	4.37 ± 0.824	4.29 ± 0.782	15.478	**0.000**	E Vs S, W Vs S, N Vs S
5.11	4.25 ± 0.905	3.97 ± 0.957	4.30 ± 0.888	4.38 ± 0.825	6.789	**0.000**	S Vs E, S Vs W, S Vs N

The following table, **Table Q5.B** summarises the results of one-way ANOVA test to find out any inter-zone relationship for the Teacher group with respect to their responses to all items in the section. As the table suggests, statistically no significant differences exist between the mean scores for the teachers of different zones for all items in section-5 except Q5.1, Q5.5, Q5.6 and Q5.7. Also, Post Hoc test (LSD) reveals the significant pairs in terms of zones. In the table, S stands for South, W stands for West, N stands for North and E stands for East.

Table Q5.B-Teachers

Item	Mean ± SD				Inter Zone Comparison One Way ANOVA		
	East (n=44)	South (n=85)	West (n=66)	North (n=183)	F	p	Post Hoc Test (LSD) Significant Pairs
5.1	3.61 ± 1.205	4.07 ± 0.973	3.97 ± 0.960	4.08 ± 0.986	2.672	**0.047**	E Vs S, E Vs N
5.2	4.16 ± 0.608	4.34 ± 0.716	4.39 ± 0.653	4.31 ± 0.801	0.950	0.417	-
5.3	4.02 ± 0.902	4.26 ± 0.833	4.30 ± 0.656	4.32 ± 0.784	1.751	0.156	-
5.4	3.48 ± 0.976	3.46 ± 1.249	3.44 ± 1.291	3.56 ± 1.202	0.230	0.876	-
5.5	3.45 ± 1.109	4.00 ± 0.988	4.00 ± 1.095	3.80 ± 1.092	3.089	**0.027**	E Vs S, E Vs W
5.6	3.27 ± 1.208	3.71 ± 1.121	3.83 ± 1.117	3.81 ± 1.118	2.939	**0.033**	E Vs S, E Vs W, E Vs N

5.7	3.43 ± 1.021	3.82 ± 1.026	3.92 ± 1.181	3.97 ± 0.931	3.446	**0.017**	E Vs W, E Vs S, E Vs N
5.8	3.80 ± 0.734	4.00 ± 0.939	3.77 ± 1.148	3.96 ± 0.974	1.043	0.373	-
5.9	4.07 ± 0.334	4.35 ± 0.735	4.08 ± 0.900	4.19 ± 0.799	2.143	0.094	-
5.10	4.09 ± 0.676	4.22 ± 0.807	4.15 ± 0.996	4.13 ± 0.861	0.309	0.819	-
5.11	4.11 ± 0.722	4.20 ± 0.737	4.08 ± 0.900	4.15 ± 0.831	0.318	0.812	-

Observations Related to Section-6 of the Questionnaire:

Section-6 of the questionnaire was related to the problem area **"Entrepreneurship /Business opportunities after the completion of BAMS course".** There were a total of six items in this section. Items are serially numbered from Q6.1 to Q6.6. Following tables summarise the responses of both the groups.

Q6.1

Students are not trained in basic management skills required to launch a new Ayurvedic hospital/ Panchakarma center/ Ayurvedic Pharmacy during BAMS course.

Table Q6.1

		SD	D	U	A	SA	Mean± SD	t	p
Students	N	18	63	11	246	306	4.18 ± 1.051	2.170	**0.030**
	%	2.8%	9.8%	1.7%	38.2%	47.5%			
Teachers	N	5	44	13	187	129	4.03 ± 0.981		
	%	1.3%	11.6%	3.4%	49.5%	34.1%			
Total	N	23	107	24	433	435			
	%	2.3%	10.5%	2.3%	42.4%	42.6%			

Table Q6.1 indicates that more than 85% of students and more than 80% of teachers are in the zone of agreement. Mean scores for both groups are more than 4 indicating a strong tendency of both the groups towards agreement. However, the mean scores of student group are significantly more than the mean scores of teacher group (p< 0.05), indicating a stronger tendency of student group towards agreement. Therefore, it can be said that a very significant number of teachers and students tend to agree that students are not trained in basic management skills required to launch a new Ayurvedic hospital/ Panchakarma center/ Ayurvedic Pharmacy during BAMS course.

Q6.2

Students are not exposed to the basics of economical aspects related to healthcare sector during BAMS course.

Table Q6.2

		SD	D	U	A	SA	Mean± SD	t	p
Students	N	4	54	21	307	258	4.18 ± 0.890	2.078	**0.038**
	%	0.6%	8.4%	3.3%	47.7%	40.1%			
Teachers	N	3	31	16	217	111	4.06 ± 0.856		
	%	0.8%	8.2%	4.2%	57.4%	29.4%			
Total	N	7	85	37	524	369			
	%	0.7%	8.3%	3.6%	51.3%	36.1%			

Table Q6.2 indicates that more than 85% of students and teachers are in the zone of agreement. Mean scores for both groups are more than 4 indicating a strong tendency of both the groups towards agreement. However, the mean scores of student group are significantly more than the mean scores of teacher group (p< 0.05), indicating a stronger tendency of student group towards agreement. Therefore, it can be said that a very significant number of teachers and students tend to agree that students are not exposed to the basics of economical aspects related to healthcare sector during BAMS course.

Q6.3
Most of the BAMS graduates prefer either studying PG course or they go for private practice and therefore, inspiring examples of industrially successful BAMS graduates are very few.

Table Q6.3

		SD	D	U	A	SA	Mean± SD	t	p
Students	N	1	27	18	277	321	4.38 ± 0.750	5.693	0.000
	%	0.2%	4.2%	2.8%	43.0%	49.8%			
Teachers	N	6	20	21	217	114	4.09 ± 0.840		
	%	1.6%	5.3%	5.6%	57.4%	30.2%			
Total	N	7	47	39	494	435			
	%	0.7%	4.6%	3.8%	48.3%	42.6%			

Table Q6.3 indicates that more than 90% of students and more than 85% of teachers in the present study are in the zone of agreement. Mean scores for both groups are more than 4 indicating a strong tendency of both the groups towards agreement. However, the mean scores of student group are significantly more than the mean scores of teacher group (p< 0.001), indicating a stronger tendency of student group towards agreement. Therefore, it can be said that a very significant number of teachers and students tend to agree that inspiring examples of industrially successful BAMS graduates are very few because most of the BAMS graduates prefer either studying PG course or they go for private practice.

Q6.4
Students are not introduced to the basic skills related to the management of Health tourism and emerging opportunities in this field during BAMS course.

Table Q6.4

		SD	D	U	A	SA	Mean± SD	t	p
Students	N	4	26	21	280	313	4.35 ± 0.782	2.452	0.014
	%	0.6%	4.0%	3.3%	43.5%	48.6%			
Teachers	N	2	19	11	204	142	4.23 ± 0.776		
	%	0.5%	5.0%	2.9%	54.0%	37.6%			
Total	N	6	45	32	484	455			
	%	0.6%	4.4%	3.1%	47.4%	44.5%			

Table Q6.4 is indicating that more than 90% of students and teachers are in the zone of agreement. This indicates that a very significant number of students and teachers in the study tend to agree with the statement that students are not introduced to the basic skills related to the management of Health-tourism and emerging opportunities in this field during BAMS course. Furthermore, the mean scores for student group are significantly higher than mean scores for teacher group (p<0.05).

Q6.5
Students are not exposed to the basic agricultural and marketing aspects of medicinal plants making it difficult to go for cultivation / marketing of medicinal plants.

Table Q6.5

		SD	D	U	A	SA	Mean± SD	t	p
Students	N	2	12	29	294	307	4.39 ± 0.690	3.430	**0.001**
	%	0.3%	1.9%	4.5%	45.7%	47.7%			
Teachers	N	3	17	19	193	146	4.22 ± 0.800		
	%	0.8%	4.5%	5.0%	51.1%	38.6%			
Total	N	5	29	48	487	453			
	%	0.5%	2.8%	4.7%	47.7%	44.3%			

Table Q6.5 indicates that more than 90% of students and more than 85% of teachers in the present study are in the zone of agreement. Mean scores for both groups are more than 4 indicating a strong tendency of both the groups towards agreement. However, the mean scores of student group are significantly more than the mean scores of teacher group ($p < 0.005$), indicating a stronger tendency of student group towards agreement. Therefore, it can be said that a very significant number of teachers and students tend to agree that students are not exposed to the basic agricultural and marketing aspects of medicinal plants during BAMS course making it difficult to go for cultivation / marketing of medicinal plants.

Q6.6
Students are not exposed to the basic manufacturing techniques related to cosmetic products and such other popular dosage forms during BAMS course making them unfit for modern pharmaceutical industry.

Table Q6.6

		SD	D	U	A	SA	Mean± SD	t	p
Students	N	4	42	20	246	332	4.34 ± 0.867	3.102	**0.002**
	%	0.6%	6.5%	3.1%	38.2%	51.6%			
Teachers	N	3	28	13	195	139	4.16 ± 0.864		
	%	0.8%	7.4%	3.4%	51.6%	36.8%			
Total	N	7	70	33	441	471			
	%	0.7%	6.8%	3.2%	43.2%	46.1%			

Table Q6.6 indicates that more than 85% of students and teachers in the present study are in the zone of agreement. Mean scores for both groups are more than 4 indicating a strong tendency of both the groups towards agreement. However, the mean scores of student group are significantly more than the mean scores of teacher group ($p < 0.005$), indicating a stronger tendency of student group towards agreement. Therefore, it can be said that a very significant number of teachers and students tend to agree that students are not exposed to the basic manufacturing techniques related to cosmetic products and such other popular dosage forms during BAMS course making them unfit for modern pharmaceutical industry.

Inter Zone comparison of means by One-way ANOVA:

The following table, **Table Q6.A** summarises the results of one-way ANOVA test to find out any inter-zone relationship for the Student group with respect to their responses to all items in the section. As the table suggests, statistically highly significant differences exist between the mean scores for the students of different zones for all items in section-6. Also, Post Hoc test (LSD) reveals the significant pairs in terms of zones. In the table, S stands for South, W stands for West, N stands for North and E stands for East.

Table Q6.A-Students

Item	Mean ± S.D.				Inter Zone Comparison One way ANOVA		
	East (n=87)	South (n=145)	West (n=150)	North (n=262)	f	p	Post Hoc Test (LSD) Significant Pairs
6.1	4.29 ± 0.806	3.68 ± 1.274	4.17 ± 1.071	4.43 ± 0.867	17.523	**0.000**	N Vs W, S Vs W, S Vs N, S Vs E
6.2	4.13 ± 0.873	3.78 ± 1.108	4.29 ± 0.805	4.36 ± 0.723	15.232	**0.000**	N Vs E, S Vs E, S Vs W, S Vs N
6.3	4.37 ± 0.717	4.13 ± 0.892	4.53 ± 0.621	4.44 ± 0.713	8.108	**0.000**	S Vs W, S Vs N, S Vs E
6.4	4.33 ± 0.623	4.12 ± 0.990	4.39 ± 0.817	4.47 ± 0.641	6.745	**0.000**	S Vs W, S Vs N, S Vs E
6.5	4.30 ± 0.684	4.21 ± 0.781	4.41 ± 0.716	4.50 ± 0.599	6.162	**0.000**	E Vs N, S Vs W, S Vs N
6.6	4.34 ± 0.744	4.08 ± 1.051	4.24 ± 0.988	4.53 ± 0.653	9.309	**0.000**	S Vs E, S Vs N, N Vs W

The following table, **Table Q6.B** summarises the results of one-way ANOVA test to find out any inter-zone relationship for the Teacher group with respect to their responses to all items in the section. As the table suggests, statistically no significant differences exist between the mean scores for the teachers of different zones for any item in section-6.

Table Q6.B-Teachers

Items	Mean ± SD				Inter Zone Comparison One way ANOVA		
	East (n=44)	South (n=85)	West (n=66)	North (n=183)	f	p	Post Hoc Test (LSD) Significant Pairs
6.1	3.82 ± 1.063	4.16 ± 0.924	3.98 ± 1.030	4.04 ± 0.965	1.278	0.282	-
6.2	3.86 ± 0.955	4.18 ± 0.743	3.98 ± 0.920	4.09 ± 0.854	1.532	0.206	-
6.3	4.18 ± 0.691	4.04 ± 0.837	4.03 ± 1.022	4.12 ± 0.803	0.483	0.694	-
6.4	4.16 ± 0.776	4.24 ± 0.718	4.17 ± 0.870	4.27 ± 0.770	0.413	0.744	-
6.5	4.20 ± 0.795	4.28 ± 0.734	4.14 ± 0.926	4.23 ± 0.786	0.424	0.736	-
6.6	4.16 ± 0.776	4.12 ± 0.905	4.08 ± 0.950	4.21 ± 0.834	0.506	0.678	-

Observations for Section-7 of the Questionnaire:

The section-7 of the questionnaire contained a single item. The respondents were asked to choose the ideal medical education system for India between three choices. The statement was as follows:

Q7

In your opinion, the "Ideal system of medical education for India" would be: (Mark (√) against any one statement that you think is the best option).

1	The existing system of multiple streams of medical education should continue as such in parallel with conventional modern medical education.	
2	Only one kind of medical degree of graduate level has to be there in India. Subject content of Indian medical systems like Ayurveda, Unani, Siddha must be included in the same curriculum along with conventional modern medicine.	
3	Only one kind of medical degree of graduate level has to be there in India which should be in conventional modern medicine. Only at Post Graduate level, one should be given an option of alternative systems of medicine.	

Table Q7.A

		Option 1	Option 2	Option 3	
Students	Number	141	399	104	644
	%	21.9%	62.0%	16.1%	100.0%
Teachers	Number	85	258	35	378
	%	22.5%	68.3%	9.3%	100.0%
Total	Number	226	657	139	1022
	%	22.1%	64.3%	13.6%	100.0%

Table Q7.A indicates the number of students and teachers for each option. It is clear from the above figures that maximum number of teachers (68.3%) and students (62%) opted for the second option i.e., Only one kind of medical degree of graduate level has to be there in India. Subject content of Indian medical systems like Ayurveda, Unani, Siddha must be included in the same curriculum along with conventional modern medicine.

Table Q7.B

			Option 1	Option 2	Option 3
Status	**BAMS Students**	Number	65	186	21
		%	23.9%	68.4%	7.7%
	PG Students	Number	76	213	83
		%	20.4%	57.3%	22.3%
	Lecturers	Number	51	163	33
		%	20.6%	66.0%	13.4%
	Readers	Number	20	39	1
		%	33.3%	65.0%	1.7%
	Professors	Number	14	56	1
		%	19.7%	78.9%	1.4%
	Total	Number	226	657	139
		%	22.1%	64.3%	13.6%

Table Q7.B shows the actual number of participants according to their status under each option. This also suggests that the option No.2 was selected by maximum number of participants from each category.

Observations Related to Section-8 of the Questionnaire:

Section-8 of the questionnaire was related to the problem area **"Relevance of Ayurveda at Personal level"**. There were a total of six items in this section. Items are serially numbered from Q8.1 to Q8.6. Following tables summarise the responses of both the groups. Tables also show the mean scores and results of Independent Samples-T Test for each item in terms of 't' value and 'p' value.

Q8.1

I personally have not undergone any *'Panchakarma'* procedure during my student life to have a first hand experience.

Table Q8.1

		SD	D	U	A	SA	Mean± SD	t	p
Students	N	163	188	13	122	158	2.88 ± 1.570	-0.394	0.694
	%	25.3%	29.2%	2.0%	18.9%	24.5%			
Teachers	N	66	134	4	112	62	2.92 ± 1.416		
	%	17.5%	35.4%	1.1%	29.6%	16.4%			
Total	N	229	322	17	234	220			
	%	22.4%	31.5%	1.7%	22.9%	21.5%			

As the **Table Q8.1** indicates, the mean scores of both the groups are less than 3 suggesting a general tendency towards disagreement with the statement. Total number of teachers and students in the zone of agreement is marginally less than the total number of students and teachers in the zone of disagreement. Therefore, it can be said that generally students and teachers tend to disagree that they personally have not undergone any *'Panchakarma'* procedure during my student life to have a first hand experience.

Q8.2

I personally don't prefer Ayurvedic medicines as a first choice for all health needs of myself and my family members.

Table Q8.2

		SD	D	U	A	SA	Mean± SD	t	p
Students	N	195	246	26	110	67	2.39 ± 1.346	1.998	0.046
	%	30.3%	38.2%	4.0%	17.1%	10.4%			
Teachers	N	114	164	15	71	14	2.22 ± 1.174		
	%	30.2%	43.4%	4.0%	18.8%	3.7%			
Total	N	309	410	41	181	81			
	%	30.2%	40.1%	4.0%	17.7%	7.9%			

Table Q8.2 suggests that more than 65% of students and more than 70% of teachers are in the zone of disagreement. Therefore, it can be said that students and teachers tend to disagree that they personally don't prefer Ayurvedic medicines as a first choice for all health needs of themselves and their family members.

Q8.3

I rarely prefer *'Rasaushadhis'* for my family members and close friends because I am not convinced about their safety.

Table Q8.3

		SD	D	U	A	SA	Mean± SD	t	p
Students	N	96	194	46	212	96	3.03 ± 1.351	2.010	0.045
	%	14.9%	30.1%	7.1%	32.9%	14.9%			
Teachers	N	66	119	29	132	32	2.85 ± 1.298		
	%	17.5%	31.5%	7.7%	34.9%	8.5%			
Total	N	162	313	75	344	128			
	%	15.9%	30.6%	7.3%	33.7%	12.5%			

Table Q8.3 suggests that students and teachers are almost equally divided in the zones of agreement and disagreement for this item. However, mean score is more than 3 in case of student group indicating a marginal tendency towards agreement whereas the same in teacher group is less than 3. Therefore, students marginally tend to agree that they rarely prefer *'Rasaushadhis'* for their family members and close friends because of safety concerns. But the teachers tend to disagree with the statement. The difference in mean scores is statistically significant (p<0.5).

Q8.4

I do not follow the directives given in *Svasthavrtta* under the topics like *'Viruddhāhāra'*, *'Sadvrtta'*, *'Dinacarya'* and *'Rtucharya'* in day to day living.

Table Q8.4

		SD	D	U	A	SA	Mean± SD	t	p
Students	N	96	216	61	189	82	2.91 ± 1.315	4.119	0.000
	%	14.9%	33.5%	9.5%	29.3%	12.7%			
Teachers	N	62	177	25	88	26	2.57 ± 1.206		
	%	16.4%	46.8%	6.6%	23.3%	6.9%			
Total	N	158	393	86	277	108			
	%	15.5%	38.5%	8.4%	27.1%	10.6%			

Table Q8.4 indicates that the mean scores of both the groups are below 3 indicating a general tendency towards disagreement. The total number of students and teachers in the zone of disagreement is more than the number in the zone of agreement. Therefore, it may be said that the students and teachers in general, tend to disagree with the statement that they do not follow the directives given in *Svasthavrtta* under the topics like *'Viruddhāhāra'*, *'Sadvrtta'*, *'Dinacarya'* and *'Rtucharya'* in day to day living.

Q8.5
I cannot precisely identify most of the herbs used in Ayurvedic pharmaceutics.

Table Q8.5

		SD	D	U	A	SA	Mean± SD	t	p
Students	N	55	206	48	254	81	3.16 ± 1.240	4.371	0.000
	%	8.5%	32.0%	7.5%	39.4%	12.6%			
Teachers	N	51	147	25	134	21	2.81 ± 1.213		
	%	13.5%	38.9%	6.6%	35.4%	5.6%			
Total	N	106	353	73	388	102			
	%	10.4%	34.5%	7.1%	38.0%	10.0%			

Table Q8.5 shows that mean score for student group is more than 3 indicating a tendency towards agreement and the mean score for teacher group is less than 3 indicating a tendency towards disagreement. This difference in mean scores is statistically highly significant (p<0.001). Therefore, it may be said that students tend to agree that they can not precisely identify most of the herbs used in Ayurvedic pharmaceutics whereas; the teachers tend to disagree with the statement.

Q8.6
I am not confident of preparing simple Ayurvedic medicines sufficient enough for dispensing in my clinical practice.

Table Q8.6

		SD	D	U	A	SA	Mean± SD	t	p
Students	N	154	226	35	145	84	2.66 ± 1.393	5.527	0.000
	%	23.9%	35.1%	5.4%	22.5%	13.0%			
Teachers	N	109	175	23	55	16	2.19 ± 1.131		
	%	28.8%	46.3%	6.1%	14.6%	4.2%			
Total	N	263	401	58	200	100			
	%	25.7%	39.2%	5.7%	19.6%	9.8%			

Table Q8.6 indicates that the mean scores of both the groups are below 3 indicating a general tendency towards disagreement. The total number of students and teachers in the zone of disagreement is significantly more than the number in the zone of agreement. Therefore, it may be said that the students and teachers in general, tend to disagree with the statement that they are not confident of preparing simple Ayurvedic medicines sufficient enough for dispensing in their clinical practice.

Inter Zone comparison of means by One-way ANOVA:

The following table, **Table Q8.A** summarises the results of one-way ANOVA test to find out any inter-zone relationship for the Student group with respect to their responses to all items in the section. As the table suggests, statistically significant differences exist between the mean scores for the students of different zones for all items in section-8 except for Q8.5. Also, Post Hoc test (LSD) reveals the significant pairs in terms of zones. In the table, S stands for South, W stands for West, N stands for North and E stands for East.

Table Q8.A-Student

Item	Mean ± Std. Deviation				Inter Zone comparison One way ANOVA		
	East (n=87)	South (n=145)	West (n=150)	North (n=262)	F	p	Post Hoc Test (LSD) Significant Pairs
8.1	3.22 ± 1.543	2.57 ± 1.475	2.62 ± 1.600	3.09 ± 1.566	6.312	**0.000**	E Vs S, E Vs W, N Vs S, N Vs W
8.2	2.36 ± 1.171	1.94 ± 1.226	2.19 ± 1.333	2.77 ± 1.373	14.457	**0.000**	E Vs S, E Vs N, N Vs S, N Vs W
8.3	3.10 ± 1.312	3.03 ± 1.346	2.53 ± 1.355	3.29 ± 1.292	10.626	**0.000**	E Vs W, S Vs W, N Vs W
8.4	3.07 ± 1.328	2.83 ± 1.238	2.70 ± 1.345	3.03 ± 1.323	2.687	**0.046**	E Vs W, N Vs W
8.5	3.17 ± 1.305	3.10 ± 1.200	2.95 ± 1.287	3.29 ± 1.201	2.526	0.057	-
8.6	2.97 ± 1.528	2.19 ± 1.230	2.27 ± 1.361	3.03 ± 1.317	18.558	**0.000**	E Vs S, E Vs W, N Vs S, N Vs W

The following table, **Table Q8.B** summarises the results of one-way ANOVA test to find out any inter-zone relationship for the Teacher group with respect to their responses to all items in the section-8. As the table suggests, except for the item Q8.3, no statistically significant differences exist between the mean scores for teachers of different zones for items in this section. Also, Post Hoc test (LSD) reveals the significant pairs in terms of zones.

Table Q8.B-Teacher

Item	Mean ± Std. Deviation				Inter Zone comparison One way ANOVA		
	East (n=44)	South (n=85)	West (n=66)	North (n=183)	F	p	Significant Pairs
8.1	3.00 ± 1.258	2.55 ± 1.358	2.97 ± 1.392	3.05 ± 1.467	2.562	0.055	-
8.2	2.18 ± 1.126	2.04 ± 1.128	2.24 ± 1.241	2.32 ± 1.180	1.139	0.333	-
8.3	2.57 ± 1.265	2.82 ± 1.399	3.29 ± 1.356	2.78 ± 1.207	3.442	**0.017**	E Vs W, S Vs W, N Vs W
8.4	2.59 ± 1.282	2.60 ± 1.157	2.74 ± 1.328	2.50 ± 1.167	0.690	0.558	-
8.5	2.68 ± 1.157	2.68 ± 1.147	3.05 ± 1.306	2.81 ± 1.219	1.308	0.271	-
8.6	2.07 ± 0.998	2.04 ± 1.063	2.45 ± 1.372	2.20 ± 1.087	1.920	0.126	-

Discussion

In India, although western or modern medicine is well developed and is liberally used, the population depends to a significant extent on traditional medicine for its health needs. India is a real example of medical pluralism where as many as six systems of medicine are officially practiced in addition to the main stream medicine i.e. Allopathy. All systems have their independent working infrastructure in the sector of medical education, practice and research. Ayurveda is the major alternative system of medicine in India.

During last 200 years, trends of institutionalisation began in the field of Ayurveda education. This trend took over a century to reach the present status of institutionalisation and university level Education. Today there are more than 200 full-fledged Ayurveda Colleges spread all over the country affiliated to different leading Universities. With the growing institutionalisation of education in Ayurveda, need has been felt to launch research and development activities in order to update it in terms of its understanding and application to the present day needs of the people. Although there are controversies in India about the approach and methodology of research in Ayurveda and about the quantum of utilisation of western modem sciences for this purpose, consensus is in favour of utilising all possible aids of modem science and technology to investigate the problems of Ayurveda and to generate evidence for the safety ad efficacy of its medications. Ayurveda today is an official system of medicine undergoing fast revival and development on scientific lines towards the need of the nation. (Singh RH, 2003).

In the last century, the education and practice of Ayurveda has passed through a fast transition from Guru-Śishya tradition to an institutionalised training system and hospital/dispensary based practice both in private and public sector. The present estimated number of registered traditional practitioners in India is over 5,00,000. Probably this number is higher than the number of conventional modem doctors in the country. Such practitioners form a huge professional manpower available within the country. If given support and provided time to time training and supplied working facilities, they could prove a big help in national health care delivery system for the masses. This is a developing country which is still facing multifaceted health problems (Singh RH, 2003).

According to Prof. R. H. Singh, (Singh RH, 2003) Ayurveda and other indigenous systems of medicine have two-fold contemporary strength which needs to be utilised:

1. In primary health-care, Ayurveda has strength related to promotive and preventive Health Care through its time-honoured lifestyle measures, dietic regimen and restorative and rejuvenative remedies of *Rasāyana Tantra*.

2. In tertiary care of chronic intractable diseases and degenerative disorders, where conventional modem medicine has not much to offer. These two are the important sectors of health care delivery system today, and in both these sectors Ayurveda and Indigenous Systems of Medicine have real strength.

Recently, with the establishment of a separate department of Indigenous Systems of Medicine (ISM) in the Ministry of Health, Government of India, the endeavour to support the growth and to maximise the use of ISM has started yielding results. Integration of ISM with conventional modem medicine and its mainstreaming is being attempted at all levels. Attempts are being made to start service facilities of Traditional Medicine specially Ayurveda in major modem medicine hospitals too. A course capsule is being developed to introduce Ayurveda in the formal curriculum of graduate studies of modem medicine in conventional modern medical colleges. The Ayurvedic Colleges have already incorporated Basic Modem Medical Sciences and diagnostics in the Syllabi of Ayurvedic Graduate Courses. Several

Overseas Universities particularly US Universities have shown interest to send their medical students for elective training in Ayurveda. All this is gradually bringing Ayurveda in the main stream of medicine and is helping its integration. It is hoped that this trend of integration and interdisciplinary development will globalise Ayurveda.

However, the problem which comes in the way of globalisation of Ayurveda is its distinct uniqueness and the ideological and the linguistic differences from the main stream of medical sciences of today (Singh RH, 2003). Furthermore, there has been a growing trend of privatization of medical education in India in recent years. Also, there has been a mushroom-growth of Ayurvedic colleges with poor infrastructural facilities (Joshi VK, 2003). Good infrastructure in terms of spacious and equipped class rooms, laboratories, well-equipped departments, hospital with well equipped labour room, operation theatre and *Pancakarma* facilities are not available in many of the Ayurvedic education institutions. Poor monitoring of these institutions by governing bodies and liberal licensing has been a cause of concern (Dwivedi M and Gupta SN, 2008).

The following observations of the present study are worth discussing in relation to specific problem-areas:

1. Extent of exposure to the basic clinical skills during BAMS Course

A very significant number of participants tend to agree that students are not trained to handle the clinical emergencies of primary healthcare level through Ayurvedic methods during BAMS course. Also, there is a general tendency towards agreement that students are not exposed to any successful Ayurvedic method of primary healthcare in the management of infectious conditions during BAMS course. A significant number of participants in the study also tend to agree that students are not exposed to any successful Ayurvedic method of primary healthcare in the management of poisoning during BAMS course.

These observations indicate that BAMS graduates are generally not trained to handle clinical emergencies of primary healthcare level through Ayurvedic methods. Also, they are not trained to handle infectious conditions through Ayurvedic methods. Even though exposure to emergency medicine has been made mandatory in the curriculum of BAMS, the students are not exposed sufficiently to this area. Probably this is the reason why BAMS graduates tend to practice Allopathy in their clinics. In India, a considerable amount of primary healthcare is delivered by BAMS graduates, especially in rural setup. Moreover, the incidence of infectious diseases is very high. This situation probably compels a BAMS graduate to practice Allopathy, though legally one can not do so in most of the states.

The study shows that there is a tendency towards agreement that students are not exposed sufficiently to the basic clinical skills and procedures like incision and drainage, suturing and catheterization during BAMS course. Also, students in general, tend to agree more strongly with the statement than teachers. But, interestingly, students and teachers from East zone tend to disagree with the statement.

This observation shows that the extent of clinical exposure to certain basic procedures and techniques is poor in Ayurvedic educational institutions. These procedures are system- neutral and a physician of any stream is generally expected to be skilful in these areas. This is especially true in primary healthcare delivery system. Probably the lack of required infrastructure like equipments and instruments in the institutions is the reason for this state of affairs.

As per the present study, there is a tendency towards agreement that students are not trained sufficiently to conduct normal delivery during BAMS course. Also, students in general, tend to agree more strongly with the statement than teachers.

In fact, there is exhaustive literature in Ayurveda regarding obstetrics. Also, conducting a normal delivery usually does not require high level of Allopathic interventions. Even then, BAMS graduates are generally not trained sufficiently in this area. In primary healthcare delivery system, especially in rural areas, BAMS graduates may prove themselves useful if they are trained in basics of obstetrics. Lack of infrastructure like labour room, lack of trained nursing staff and lack of skilful doctors in Ayurvedic education institutions may be the reason for poor training in this field.

In the current study, a general tendency towards agreement is noted with the statement that students are not exposed to a large variety of cases during BAMS course because patients visiting Ayurvedic institutions belong to only few identifiable categories.

This observation indicates that patients visiting Ayurvedic education institutions belong to only few identifiable categories. Therefore, it is the need of the hour that new target disease categories are identified and new management modalities are explored by the Ayurvedic community. Also, the potential of Ayurvedic system of medicine needs to be explained to the society. Furthermore, the Government needs to involve Ayurvedic physicians in national healthcare programmes so that Ayurveda is brought into the mainstream.

As the study indicates, there is a general tendency towards agreement that students are not exposed sufficiently to the basic modern knowledge of the subjects like Physiology, Pathology, Biochemistry, Pharmacology, Medicine, Paediatrics, Obstetrics & Gynaecology, Eye & ENT and Surgery during BAMS course. Furthermore, students tend to agree with this statement more strongly than the teachers. However, teachers from East zone tend to disagree with the statement.

Basic modern knowledge in the subjects cited above is essential in the present era. This basic knowledge is required to understand the investigation reports, to evaluate the prognosis and to communicate the Ayurvedic concepts in terms of modern knowledge to those who don't know Ayurveda. Furthermore, basic knowledge in these areas minimises the prescription errors due to ignorance. For example, one who knows the implications of liver diseases will not prescribe a herb or Ayurvedic preparation which might have hepatotoxicity associated with its usage. Also, a basic understanding in these subjects makes a physician conversant with professionals of other streams of Medicine. Furthermore, the basic knowledge in these subjects enables one to make contributions in research areas too.

The reasons for insufficient training in these areas could be many: 1. No trained teachers are probably available in Ayurvedic education institutions who can train students in these areas. This may be because of the deficiencies in post graduate education as far as basic contemporary knowledge is concerned. 2. No teachers from modern medical colleges are involved in Ayurvedic education because of policy-related issues. To solve this problem, some changes in education policy are required. An integrated model of education system may be the ideal answer to this problem, where indigenous medicine is also incorporated in the mainstream medical education and practice.

As the present study indicates, there is a general tendency towards agreement that students are not exposed sufficiently to the basic skills of interpreting ECG, X-Ray and such other diagnostic tools and their clinical utility during BAMS course. Furthermore, students tend to agree with the statement more strongly than teachers.

Understanding the basics of modern investigation methods is essential to a medical graduate of any stream because this helps in arriving at a definite diagnosis. Diagnosis determines the line of treatment and final outcome of the treatment is in turn, dependent on right choice of the management strategy. The ancient diagnostic techniques explained in classical textbooks of Ayurveda are probably inadequate in the present era or their sufficiency needs to be scientifically established.

A very significant number of students and teachers in the present study tend to agree that students are not exposed to the basic skills in the areas like Genetic counselling, Human sexuality, End of life care, Geriatrics and Drug and alcohol abuse during BAMS course. Students tend to agree with the statement more strongly than teachers.

A BAMS graduate is required to be trained in these areas because patients often seek advice from the practitioners of alternative medicine in these somewhat complicated issues. Furthermore, a significant amount of authentic literature is available in Ayurvedic textbooks related to *Vājīkaraṇa, Rasāyana* and *Madātyaya.* Therefore, it will not be irrelevant to train Ayurvedic graduates in these areas.

Students in the present study, show a marginal tendency towards agreement that students are not exposed sufficiently to the basic methods of physical examination, diagnosis and management of common clinical conditions, making them non-confident clinicians/ practitioners.

Only a good exposure to basic clinical skills can produce a confident physician. Therefore, it is not an unusual perception from student perspective. Furthermore, these observations clearly indicate that there are some serious flaws in the existing system of graduate level Ayurvedic education. The system has been unable to produce confident clinicians who would be useful to cater the primary healthcare needs of the country. Though, many topics related to essential clinical skills are included in the curriculum, the system has not been successful in producing good clinicians.

Probably, teaching job has become less attractive than private practice in terms of monitory earnings. This discourages good clinicians from joining Ayurvedic educational institutions. Therefore, there is a scarcity of good teachers in the system. Due to the mushrooming of private Ayurvedic colleges, the situation has even become worse. In many private institutions, teachers are exploited. There is no regulatory authority to monitor the pay packages and timely promotions of teachers in these institutions. This may be one of the reasons why the system is failing in producing good clinicians.

As the study suggests, there is a general tendency towards agreement that students are not trained sufficiently in the basic clinical methods related to *Pancakarma, Ksāra Sūtra* and *Jalaukāvacarana* during BAMS course. Furthermore, students in general, tend to agree with the statement more strongly than teachers.

This observation is discouraging because the students are not being trained at a desired level in unique and specific skills related to Ayurvedic system of medicine. One of the reasons for this may be that the number of patients visiting Ayurvedic institutions may be comparatively less and students are not exposed to a wide variety of cases. Also, wherever there is Post Graduate education, these special skills are probably being imparted to Post Graduate scholars only. Therefore, BAMS scholars are being deprived of these skills. This is a serious issue that has to be given a serious thought.

In this regard, there are several questions that need to be answered:

Are these skills so special that BAMS graduates should not be trained to practice them?

Is highly specialised training in these areas really given to PG scholars?

If not, what is the logic behind depriving BAMS graduates from acquiring these skills?

Finally, a clear policy needs to be developed which should address these issues.

II. Job opportunities after the completion of BAMS course

A significant number of participants in the present study tend to agree that in most of the states, a BAMS degree holder can not practice Allopathy legally and therefore, hospitals generally prefer MBBS graduates as medical officers instead of BAMS graduates. Furthermore, students tend to agree with the statement more strongly than teachers. They also tend to agree that Ayurvedic hospitals are less in number in comparison to Allopathic ones and therefore, job opportunities are limited for a BAMS graduate. Furthermore, a significant number of participants in the present study tend to agree with the statement that in Ayurvedic educational institutions, only Post Graduate doctors are employed and not BAMS degree holders. The participants also tend to think that job opportunities are limited for BAMS graduates in research institutions. There is a general tendency towards agreement that even in Government sector, BAMS graduates are not treated at par with MBBS graduates and therefore, job opportunities are limited in certain areas e.g., Railways and defence. Also, as the study suggests, there is a general tendency towards agreement that Ayurvedic pharmaceutical firms prefer Post Graduate candidates instead of BAMS degree holders as experts. Participants also tend to agree that there is lot of competition for jobs among BAMS degree holders as a result of mushrooming of Ayurvedic colleges. Students in the present study consider all these problems to be more serious ones than teachers.

These observations indicate that there is a real problem related to the job opportunities for BAMS graduates. Also, the study indicates that there is a considerable amount of anxiety related to career-opportunities among students. This anxiety is probably considerably less among teachers because they are already into a job. This may be the reason why students perceive this problem to be a more serious one. Furthermore, Government is required to look into the matter related to creation of job opportunities for BAMS graduates in certain departments like Railways and Defence. In teaching institutions too, some posts like tutors and medical officers may be created for BAMS graduates. In Indian Administrative Services too, Ayurveda needs to be included as an optional subject just like modern medicine. If the quality of education is improved, some job opportunities may open up in research institutes and in other hospitals. Furthermore, a clear policy regarding the eligibility of BAMS graduates to practice allopathy needs to be evolved.

III. Scientific relevance of the Curriculum of BAMS course

Participants in the current study tend to agree that most of the topics in the subject *'Ayurvedīya Itihāsa'* have least practical applicability. Furthermore, students tend to agree more strongly with the statement than teachers.

The solution for this problem could be to remove the separate paper titled *'Ayurvedīya Itihāsa'* from the BAMS curriculum and to introduce the relevant historical topics in the respective subjects. Furthermore, only those topics having practical applicability need to be retained in the curriculum.

Participants in the present study generally tend to agree that most of the topics covered in the subject *'Padārtha Vijnyāna'* are philosophical and their practical applicability is limited. Also, students tend to agree more strongly with the statement than teachers.

This observation simply means that the education system has been unable to convince the students regarding the applicability of certain basic concepts of Ayurveda.

The participants in the present study generally tend to agree that many topics in *'Rachanā Śārīra'* like *'Marma'*, *'Sirā'*, *'Snāyu'*, *'Sandhi'* etc. are outdated as more advanced knowledge on these topics is available in the textbooks of Modern Anatomy/ Modern surgery. Also, students in general tend to agree more strongly with the statement than teachers. Interestingly, however, the students from West zone tend to disagree with the statement.

Participants in general tend to agree that topics like 'Assessment of *Prakrti* and *Dhātu Sāra*' are given undue importance in the subject *'Kriyā Śārīra'* and the clinical applicability of these topics is not emphasized in clinical disciplines. Also, students tend to agree more strongly with the statement than teachers.

Basic theories and applied aspects of any stream of science have to go hand in hand. Undue emphasis on theoretical basics and negligence on applied aspects discourages the students from applying the basic concepts clinically. Therefore, a proportionate and reasonable importance has to be given in the curriculum on both theory and practical aspects. As the study indicates, this re-arrangement has to be made in the curriculum of BAMS course. Topics like *'Prakrti'* should receive equal emphasis in clinical disciplines as in the subject *'Kriyā Śārīra'*.

A significant proportion of participants tend to agree that the essential practical exposure to laboratory diagnostic methods in serology, immunology, histopathology, microbiology and parasitology is not emphasized in *'Roga nidāna* and *Vikrti Vijñāna'*. Also, students tend to agree more strongly with the statement than teachers.

Though, most of these topics are included in the BAMS curriculum, sufficient exposure to these basic diagnostic methods is not provided during the BAMS course. Lack of infrastructural facilities and technically trained personnel in Ayurvedic institutions might be the cause for this state of affairs.

A significant proportion of participants in the present study tend to agree with the statement that in *'Dravyaguna'* essential basic information related to recent advances in pharmacodynamic/ pharmacognostic/ phytochemical attributes of various Ayurvedic herbs and methods of evaluation of their pharmacological effects is not emphasized. Also, students tend to agree more strongly with the statement than teachers.

Participants in the present study generally tend to agree that essential basic knowledge related to various technologically advanced methods of 'Drug Standardization' is not included in the curricula of either *'Dravyaguna'* or *'Rasa Śāstra'*. Also, students tend to agree more strongly with the statement than teachers.

Most of the teachers and students in the present study feel that essential basic knowledge related to pharmaco-vigilance, safety profile, toxicity studies and Good Manufacturing Practices should be included in *'Rasa Śāstra'*.

Participants also tend to agree that essential basic knowledge related to the methods of quantitative and qualitative analysis of chemical components of Ayurvedic preparations is not included in the curriculum of *'Rasa Śāstra'*.

A significant number of participants tend to agree that in *'Agada Tantra'*, most of the Ayurvedic topics describing the classifications/numbers/varieties of poisons and their effects are outdated and impractical.

Significant proportion of participants in the present study tend to agree that topics related to *'Arishta Vijñāna'* explained in *'Indriya Sthāna'* of *'Caraka Samhitā'* are practically not useful because they do not fit in to the present social scenario. Students tend to agree with the statement more strongly than teachers. Interestingly, teachers from South zone tend to disagree with the statement.

Most of the participants in the present study tend to agree that practical training related to the basics of medical jurisprudence, toxicology and forensic medicine is not emphasized in teaching making a BAMS graduate inefficient in handling the legal procedures.

Significant proportion of participants in the present study tend to agree that essential information on recent studies/ reports related to efficacy of Ayurvedic medicines/ procedures is not included in the curriculum of clinical disciplines, which is required to be included. Also, students tend to agree more strongly with this statement than teachers.

A significant proportion of the participants tend to agree that the curricula of clinical disciplines contain many outdated methods of treatment/management which are impractical (e.g., *Dronī Prāveśika Rasāyana*). Also, students tend to agree more strongly with the statement than teachers.

Participants in the present study tend to agree that certain essential topics like Clinical decision making, Care of hospitalized patients, Patient interviewing skills, Ethical decision making and Geriatric patient care are not emphasized in the curricula of the clinical disciplines, which are required to be included. Also, students tend to agree more strongly with the statement than teachers.

Participants in the study generally tend to agree that certain essential topics related to medical practice including Cost effective medical practice, Quality assurance in medicine, Practice management and Medical record-keeping are not included in the curriculum, which are required to be included.

More than 70% of students and teachers in the study tend to agree that many modern technical terms are translated into 'Sanskrit' in the curriculum (e.g., *'Unduka Puccha Śotha'* for Appendicitis, *'Hrtkāryacakra'* for cardiac cycle etc.) which does not serve any practical purpose.

More than 75% of students and teachers tend to agree that many controversial topics (e.g., certain structures in *Racanā Śārīra*, certain herbs in *Dravyaguna*) are included in the curricula which lead only to confusion among students.

The above observations indicate that the curricula of BAMS course require a radical change. From classical textbook-oriented and literature- oriented learning, learning has to become more clinically oriented. The syllabi of all the subjects have to be radically reviewed and restructured. Problem- based teaching has to be adopted in Ayurveda too. Impractical and redundant topics are to be removed from the curricula. Also, controversial topics need to be removed. Translation of modern technical terms into 'Sanskrit' doesn't enrich Ayurveda. Therefore, such practice has to be stopped. Wherever the modern knowledge is to be given, it has to be in a direct way and not in a camouflaged way.

The study suggests that recent advances in technology/ research related to medicinal herbs are to be incorporated into the curriculum. Certain essential topics related to medical practice including Cost effective medical practice, Quality assurance in medicine, Practice management and Medical record-keeping are to be incorporated. Essential topics like Clinical decision making, Care of hospitalized patients, Patient interviewing skills, Ethical decision making and Geriatric patient care are required to be included. Essential basic knowledge related to pharmaco-vigilance, safety profile, toxicity studies and Good Manufacturing Practices are to be included. Essential basic knowledge related to the methods of quantitative and qualitative analysis of chemical components of Ayurvedic preparations is to be included. Essential basic information related to recent advances in pharmacodynamic/ pharmacognostic/ phytochemical attributes of various Ayurvedic herbs and methods of evaluation of their pharmacological effects have to be included. Basic knowledge related to various technologically advanced methods of 'Drug Standardization' is to be included in the curriculum.

Furthermore, mere inclusion of these topics may not be sufficient to achieve the goal. Rather, there has to be a multi-disciplinary approach in education system. Ideally, some teachers from analytical chemistry, phyto- chemistry, botany, pharmaceutical chemistry may have to be appointed in teaching institutions to impart the required training. This may prove helpful in conducting some research activities at the institution-level too.

Often comparisons are drawn between allopathic and Ayurvedic systems of education. Basically, it is to be realised that the needs of two systems are different. Therefore, as per need, many changes need to be incorporated in the curriculum of BAMS course irrespective of the fact that these changes may not be relevant to modern conventional (Allopathic) medical education.

IV. Teaching methodology that is followed in existing system of Ayurvedic education.

A significant number of students and teachers in the present study tend to agree that Ayurvedic teaching methodology does not keep up the scientific values and scientific spirit of a young student. Also, they tend to agree that the teaching methodology followed in Ayurvedic educational institutions does not encourage questioning among the young students.

It is the general perception that the interpretation of theories like *'Tridosa'* or *'Pancha Mahābhūta'* varies largely from one teacher to another making these theories further confusing and vague.

A significant proportion of students tend to agree that memorizing the classical 'Sanskrit' verses is unduly emphasized in Ayurvedic method of teaching, making the process of learning difficult.

A significant proportion of teachers and students tend to agree that memorizing the reference number of a particular chapter/ verse of any *'Samhitā'* does not serve any practical purpose, but is given undue importance in teaching and examination system.

86.9% of students and 70.3% of teachers in the study tend to agree that the examination system in Ayurveda does not assess the actual abilities and skills of a student; rather, it largely depends on the assessment of memorizing capacity of students.

A significant proportion of the participants in the study tend to agree that standard textbooks covering all the topics in curricula are not available in the market, making understanding on the subjects more difficult.

There is a general tendency towards agreement that memorizing the numbers of various structures / their measurements / various classifications of diseases *(Sankhyā Samprāpti)* as per different authors etc. does not serve any practical purpose, but is given undue importance in teaching and examination system.

A significant proportion of participants tend to agree that students are not trained in areas like using a computer-based clinical record keeping program, carrying out reasonably sophisticated searches of medical information databases on internet, using a variety of forms of telemedicine etc., making them technologically inferior.

There is a general tendency toward agreement that BAMS students are not introduced to essential 'Communication skills' like discussing a prescription error with the patient, providing safe sex counselling, negotiating with a patient who requests unnecessary investigations etc.

Above observations indicate that certain radical changes are required to be adopted in the teaching methodology. From memory –oriented teaching, teaching has to become understanding - oriented. New methods of teaching and learning have to be incorporated. Memorizing the original 'Sanskrit' verses and references from textbooks without emphasising the practical utility of doing the same should be discouraged. Only clinically/ practically applicable material has to be memorized.

As per the present study, the examination system too requires a radical change. Many students during the discussions with the investigator, have opined that six-monthly semester type of examination system be introduced in Ayurvedic education. This may reduce the burden on students. Memory- oriented questions have to be limited to certain extent in the examinations. Understanding and application –

oriented questions need to be increased. In practical / oral examinations too, memory –oriented questions have to be reduced and actual skills of a student need to be assessed. Questions such as different classifications as per different authors, recitation of verses, references related to chapter –numbers etc are required to be reduced. Even in common entrance tests that are conducted at various universities for admission in post-graduate courses, these changes are required to be introduced. Undue importance often given to 'Sanskrit' may have to be reduced in examination system also.

Problem- based learning has to be emphasised in education. This existed during ancient times. For example, while describing the pathology and clinical manifestations of the disease called *'Grahanī'*, Caraka has described the physiology of digestion and metabolism also.

Newer technological innovations and their applicability in practice of medicine are required to be introduced in the education system.

Classical textbooks need to be interpreted under the light of recent advances in medicine. Standard textbooks are to be made available. For this purpose, expert committees may be framed for each subject to prepare standard textbooks. The same textbooks need to be followed in all institutions. In recent times, authoring a textbook has become a commercial activity rather than academic one. Too many sub-standard textbooks are available in the market and students are confused in this regard.

Essential 'Communication skills' like discussing a prescription error with the patient, providing safe sex counselling, negotiating with a patient who requests unnecessary investigations etc have to be included in the education system. Students need to be trained in areas like using a computer-based clinical record keeping program, carrying out reasonably sophisticated searches of medical information databases on internet, using a variety of forms of telemedicine etc. This training may be given during internship also.

Furthermore, questioning needs to be encouraged among students. Ayurvedic concepts need to be interpreted in uniform manner in all institutions to reduce confusion among students.

V. Global Challenges being faced by the Ayurvedic system of medicine.

More than 85% of students and more than 80% of teachers in the present study tend to agree that serious questions being raised on the safety profile of Ayurvedic preparations are posing a threat to the Ayurvedic system of Medicine. A very significant proportion of participants tend to perceive the issue of standardization of Ayurvedic preparations to be still a problem that needs to be addressed. These observations indicate that the questions related to safety profile of Ayurvedic preparations are posing a threat to the Ayurvedic system of Medicine. This issue requires immediate attention. Issue of standardization of Ayurvedic preparations is still a problem that needs to be addressed. If some basic information related to the assessment of toxicity / safety and methods of standardisation is introduced in the BAMS curriculum, students may become aware of these issues.

More than 85% of students and teachers tend to agree that in many countries, practicing Ayurveda is not allowed legally and therefore, there are no opportunities for BAMS graduates in such countries. This is a global issue that needs to be addressed. Government has to make efforts in obtaining legal recognition for Ayurveda as an independent system of medicine globally.

Also, as the study shows, there is a general tendency towards agreement that Ayurvedic academicians do not figure anywhere in authoring the scientific and evidence based papers in reputed international journals and they do not voluntarily participate in International platforms to present their research data.

There is a general perception that Ayurvedic academicians do not follow international standards while planning the protocols of research projects and while writing research reports.

This indicates that Ayurvedic academicians need to volunteer in authoring the scientific and evidence based papers in reputed international journals. They are required to be trained for this purpose. They are also to be encouraged to participate in debates related to Ayurveda on international platforms. Training programmes are required to be introduced for Ayurvedic academicians where, training is to be given in preparing the research projects, planning the research protocols and in other various areas of research methodology.

Participants in the present study, generally tend to agree that Ayurvedic scholars generally do not have knowledge regarding 'Intellectual Property Rights' and patenting procedures. Some essential information related to these topics may be introduced in the curriculum making BAMS graduates conversant in these topics.

More than 80% of students and teachers in the present study tend to agree that authentic websites providing up-to-date knowledge in Ayurveda are not hosted by Ayurvedic institutions. The education institutions are required to be encouraged to host authentic websites giving information related to Ayurveda. Classical textbooks of Ayurveda may be made available online at these websites. Government also may take initiative in this regard and launch authentic websites.

A significant number of participants in the study tend to agree that no standard international indexed and peer-reviewed journals are published by Ayurvedic institutions making it difficult for Ayurvedic researches have global attention. This is a serious issue that needs immediate attention. Education institutes are required to be encouraged to pursue some research activities including publication of peer – reviewed journals.

More than 85% of students and teachers tend to agree that Pharmacodynamic/ pharmacokinetic properties/ efficacy/ safety profiles and chemical compositions of Ayurvedic formulations are yet to be established making it difficult for experts in conventional medicine to accept Ayurveda.

VI. Entrepreneurship /Business opportunities after the completion of BAMS course.

More than 80% of students and teachers in the study tend to agree that students are not trained in basic management skills required to launch a new Ayurvedic hospital/ Panchakarma center/ Ayurvedic Pharmacy during BAMS course. More than 85% of students and teachers tend to agree that students are not exposed to the basics of economical aspects related to healthcare sector during BAMS course. More than 90% of students and teachers in the study tend to agree with the statement that students are not introduced to the basic skills related to the management of Health-tourism and emerging opportunities in this field during BAMS course. More than 85% of students and teachers in the present study tend to agree that students are not exposed to the basic agricultural and marketing aspects of medicinal plants during BAMS course making it difficult to go for cultivation / marketing of medicinal plants. More than 85% of students and teachers in the present study tend to agree that students are not exposed to the basic manufacturing techniques related to cosmetic products and such other popular dosage forms during BAMS course making them unfit for modern pharmaceutical industry.

The above observations indicate that business opportunities are scarce for BAMS graduates. Some basic training is required to be given to BAMS graduates in management skills required to launch a new Ayurvedic hospital/ Panchakarma center / Ayurvedic Pharmacy. Also, basic agricultural and marketing aspects of medicinal plants related to cultivation / marketing need to be introduced at BAMS level. The basic manufacturing techniques related to cosmetic products and such other popular dosage forms are required to be introduced during BAMS course.

VII. Ideal system of medical education for India:

68.3% of teachers and 62% students in the present study tend to agree that only one kind of medical degree of graduate level has to be there in India. Subject content of Indian medical systems like Ayurveda, Unani, Siddha must be included in the same curriculum along with conventional modern medicine.

As per the perception of Ayurvedic teachers and students in the present study, total integration of medical education would be ideal for India. As per this model, only one kind of medical degree of graduate level should be there and this should include basic training in Indian Systems of Medicine also along with the training in conventional (Allopathic) medicine. This is similar to the integrated approach that is being followed in a few countries like China and Vietnam.

In India, the parallel model has proven to be deficient in many aspects as evidenced in the present study. A total integration of all medical systems will be ideal for India. New policies and laws are required to be framed to achieve this goal. At graduate level of education, all systems of medicine may be introduced and only at post graduate level, the training may be given in different systems separately.

VIII. Relevance of Ayurveda at Personal level

The present study indicates that most of the teachers and students of Ayurveda have undergone some or the other kind of *'Pancakarma'* procedure during their student life to have a firsthand experience.

The study also indicates that students and teachers of Ayurveda prefer Ayurvedic medicines as a first choice of treatment for all health needs of themselves and their family members.

The study indicates that students and teachers of Ayurveda try to follow the directives given in *'Svasthavrtta'* under the topics like *'Viruddhāhāra'*, *'Sadvrtta'*, *'Dinacarya'* and *'Rtucharya'* in day to day living.

The participants in the present study generally consider themselves to be capable of preparing simple Ayurvedic medicines sufficient enough for dispensing at their clinics.

These are positive trends and need to be looked at with a positive perspective. This indicates that students and teachers have inclination towards *'Pancakarma'* procedures and also are willing to rely on Ayurveda for their personal health needs.

Students in the present study marginally tend to agree that they rarely prefer *'Rasaushadhis'* for their family members and close friends because of safety concerns. But teachers tend to disagree with the statement. However, students from West zone tend to disagree with the statement whereas teachers from West zone marginally tend to agree with the statement.

This observation indicates that some apprehension prevails among students regarding the use of *'Rasaushadhi's* because of safety concerns.

Inter-zone Variations:

Interestingly, most of the statistically significant inter-zone variations were observed among student group only. Though, deriving any confirmatory conclusion is difficult from this observation, student community is probably more sensitive for the cultural and social influences at regional level. Furthermore, teachers in most of the institutions come from different regions and therefore regional differences are probably imperceptible. On the other hand, most of the students in these institutions are probably local people, who still retain regional identity in their attitudes.

About the Participants in the present study:

Out of 32 institutions covered in the study, 23 were of Governmental administration. Only 6 private and 3 semi-Governmental institutions were covered in the study. Also, maximum participants were from Governmental institutions (781). Therefore, this study reflects the status of Ayurvedic education system more in Governmental institutions than in private or semi-Governmental institutions.

Further, an overall **57.4%** of response rate to this study can be viewed as good considering the nature of study (mailed survey) and the lengthy questionnaire. Therefore, it may be said that there was a considerable level of willingness among the participants to participate in the study.

Summary and Conclusion

The present study was planned to evaluate the **'Relevance of current system of Ayurvedic education in the emerging global scenario'** with special reference to graduate level Ayurvedic education in India leading to a BAMS (Ayurvedacharya) degree.

The study is based on the subjective perceptions of a sample of students and teachers drawn from various Ayurvedic educational institutions spread all over India.

The study is based on a Mailed Survey, which was carried out between September 2005 and October 2008.

A methodically validated semi-structured questionnaire was used as the tool in the study. The questionnaire covered eight sections under which a total of 73 items were grouped. These sections covered the following problem areas:

1. Problems related to the exposure of a BAMS graduate to basic clinical skills
2. Problems related to job opportunities after the completion of BAMS course
3. Problems related to the relevance of the Curriculum of BAMS course
4. Problems related to Teaching methodology in the existing system of Ayurvedic education
5. Problems related to Global Challenges being faced by the Ayurvedic syestem of medicine
6. Problems related to Entrepreneurship /Business opportunities after the completion of BAMS course
7. Perception regarding the ideal system of medical education for India.
8. Problems related to Personal relevance of Ayurveda to the teachers and students.

All items in the questionnaire expressed the problems related to respective areas with a negative connotation. The respondents were given the option of recording their response in the form of 'Strongly Agree', 'Agree', 'Undecided', 'Disagree' and 'Strongly Disagree' by recording a check mark (√) in the respective columns provided for the purpose, except for Section No.7. Section No.7 contained three statements in the form of 'Multiple Choices' out of which the respondents were asked to choose only one option that they considered was the best.

Printed copies of questionnaires were mailed to about 32 Ayurvedic colleges recognised by CCIM spread all over India so that at least 10% of all institutes existing in each geographical zone were included. This process was carried out on the basis of random cluster sampling method. Also, the aim was to include as many states as possible. The heads of the institutions were requested to distribute the questionnaire among all interns/house surgeons, PG students and teachers of their institution randomly and were asked to collect them after a gap of 1-2 days.

A total of 1022 participants responded to the questionnaire. This number included 644 students and 378 teachers. The maximum number of participants was from the state of Uttar Pradesh (195) and the minimum number was from the state of Bihar (19).

Following are the salient findings of the study:

- The study indicates that BAMS graduates are generally not trained to handle clinical emergencies of primary healthcare level through Ayurvedic methods. Also, they are not trained to handle infectious conditions through Ayurvedic methods.

- The study shows the extent of clinical exposure to certain basic procedures and techniques is poor in Ayurvedic educational institutions. The study also indicates that students are not trained sufficiently to conduct normal delivery during BAMS course.

- Present study indicates that patients visiting Ayurvedic education institutions belong to only few identifiable categories. Study also suggests that students are not exposed sufficiently to the basic modern knowledge of the subjects like Physiology, Pathology, Biochemistry, Pharmacology, Medicine, Paediatrics, Obstetrics & Gynaecology, Eye & ENT and Surgery during BAMS course.

- As the present study indicates, students are not exposed sufficiently to the basic skills of interpreting ECG, X-Ray and such other diagnostic tools and their clinical utility during BAMS course.

- A very significant number of students and teachers in the present study tend to agree that students are not exposed to the basic skills in the areas like Genetic counselling, Human sexuality, End of life care, Geriatrics and Drug and alcohol abuse during BAMS course.

- As the study suggests, students are not trained sufficiently in the basic clinical methods related to *Pancakarma*, *Kshāra Sūtra* and *Jalaukāvacarana* during BAMS course.

- The study indicates that there is a real problem related to the job opportunities for BAMS graduates. Also, there is a considerable amount of anxiety related to career-opportunities among students. As the study indicates, job opportunities for BAMS graduates are poor in comparison to MBBS graduates and post graduates of Ayurveda. Also, the study suggests that BAMS graduates are not preferred for employment in education institutions and research institutions. In Government sector too, BAMS graduates are not treated at par with MBBS graduates as per the perceptions of the participants in the present study.

- The study indicates that the curricula of BAMS course require a radical change. From classical textbook oriented and literature oriented learning, learning has to become more clinically oriented. The syllabi of all the subjects have to be radically reviewed and restructured. Problem-based teaching has to be adopted in Ayurveda too. Impractical and redundant topics are to be removed from the curricula. Also, controversial topics need to be removed. The study suggests that translation of modern technical terms into 'Sanskrit' doesn't enrich Ayurveda. Therefore, such practice has to be stopped.

- The study also suggests that recent advances in technology/ research related to medicinal herbs are to be incorporated into the curriculum. Certain essential topics related to medical practice like Cost effective medical practice, Quality assurance in medicine, Practice management and Medical record-keeping are to be incorporated. Essential topics like Clinical decision making, Care of hospitalized patients, Patient interviewing skills, Ethical decision making and Geriatric patient care are required to be included. Essential basic knowledge related to pharmaco-vigilance, safety profile, toxicity studies and Good Manufacturing Practices are to be included. Essential basic knowledge related to the methods of quantitative and qualitative analysis of chemical components of Ayurvedic preparations is to be included. Essential basic information related to recent advances in pharmacodynamic/ pharmacognostic/ phytochemical attributes of various Ayurvedic herbs and methods of evaluation of their pharmacological effects have to be included. Basic knowledge related to various technologically advanced methods of 'Drug Standardization' is to be included in the curriculum.

- The study indicates that the current teaching methodology does not keep up the scientific values and scientific spirit of a young student. Also, the teaching methodology followed in Ayurvedic educational institutions does not encourage questioning among the young students.

- Study also suggests that memorizing the classical 'Sanskrit' verses is unduly emphasized in Ayurvedic method of teaching, making the process of learning difficult. A significant proportion of teachers and students tend to agree that memorizing the reference number of a particular chapter/ verse of any *Samhitā* does not serve any practical purpose, but is given undue importance in teaching and examination system. Also, as the study suggests, memorizing the numbers of various structures / measurements / classifications of diseases *(Sankhyā Samprāpti)* as per different authors etc. does not serve any practical purpose, but is given undue importance in teaching and examination system.

- A significant proportion of participants tend to agree that students are not trained in areas like using a computer-based clinical record keeping program, carrying out reasonably sophisticated searches of medical information databases on internet, using a variety of forms of telemedicine etc., making them technologically inferior.

- The study indicates that certain radical changes are required to be adopted in the teaching methodology. From memory oriented teaching, teaching has to become understanding oriented. New methods of teaching and learning have to be incorporated. Memorizing the original 'Sanskrit' verses and references from textbooks without emphasising the practical utility of doing the same should be discouraged. Only clinically/ practically applicable material has to be memorized.

- As per the present study, the examination system too requires a radical change. Memory oriented questions have to be limited to certain extent in the examinations. Understanding and application oriented questions need to be increased.

- The study indicates that newer technological innovations and their applicability in practice of medicine are required to be introduced in the education system. Standard textbooks are to be made available. Essential 'Communication skills' like discussing a prescription error with the patient, providing safe sex counselling, negotiating with a patient who requests unnecessary investigations etc have to be included in the education system.

- More than 85% of students and more than 80% of teachers in the present study tend to agree that serious questions being raised on the safety profile of Ayurvedic preparations are posing a threat to the Ayurvedic system of Medicine. A very significant proportion of participants tend to perceive the issue of standardization of Ayurvedic preparations to be still a problem that needs to be addressed. These observations indicate that the questions related to safety profile of Ayurvedic preparations are posing a threat to the Ayurvedic system of Medicine.

- Also, as the study shows, there is a general tendency towards agreement that Ayurvedic academicians do not figure anywhere in authoring the scientific and evidence based papers in reputed international journals and they do not voluntarily participate in International platforms to present their research data. There is a general perception that Ayurvedic academicians do not follow international standards while planning the protocols of research projects and while writing research reports. This indicates that Ayurvedic academicians need to volunteer in authoring the scientific and evidence based papers in reputed international journals. They are required to be trained for this purpose. They are also to be encouraged to participate in debates related to Ayurveda on international platforms. Training programmes are required to be introduced for Ayurvedic academicians where, training is to be given in preparing the research projects, planning the research protocols and in other various areas of research methodology.

- Participants in the present study generally tend to agree that Ayurvedic scholars generally do not have knowledge regarding 'Intellectual Property Rights' and patenting procedures. Some essential information related to these topics may be introduced in the curriculum making BAMS graduates conversant in these topics. More than 80% of students and teachers in the present study tend to agree that authentic websites providing up-to-date knowledge in Ayurveda are not hosted by Ayurvedic institutions. The education institutions are required to be encouraged to host authentic websites giving information related to Ayurveda. Government also may take initiative in this regard and launch authentic websites.

- A significant number of participants in the study tend to agree that no standard international indexed and peer-reviewed journals are published by Ayurvedic institutions making it difficult for Ayurvedic researches have global attention. This is a serious issue that needs immediate attention. Education institutes are required to be encouraged to pursue some research activities including publication of peer – reviewed journals. More than 85% of students and teachers tend to agree that Pharmacodynamic/ pharmacokinetic properties/ efficacy/ safety profiles and chemical compositions of Ayurvedic formulations are yet to be established making it difficult for experts in conventional medicine to accept Ayurveda.

- The study indicates that business opportunities are scarce for BAMS graduates. Some basic training is required to be given to BAMS graduates in management skills required to launch a new Ayurvedic hospital/ Panchakarma center / Ayurvedic Pharmacy. Also, basic agricultural and marketing aspects of medicinal plants related to cultivation / marketing need to be introduced at BAMS level. The basic manufacturing techniques related to cosmetic products and such other popular dosage forms are required to be introduced during BAMS course.

- 68.3% of teachers and 62% students in the present study tend to agree that only one kind of medical degree of graduate level has to be there in India. Subject content of Indian medical systems like Ayurveda, Unani, Siddha must be included in the same curriculum along with conventional modern medicine. Therefore, as per the perception of Ayurvedic teachers and students in the present study, total integration of medical education would be ideal for India. As per this model, only one kind of medical degree of graduate level should be there and this should include basic training in Indian Systems of Medicine along with the training in conventional (Allopathic) medicine. This is similar to the integrated approach that is being followed in a few countries like China and Vietnam.

- The present study indicates that most of the teachers and students of Ayurveda have undergone some or the other kind of *'Pancakarma'* procedure during their student life to have a firsthand experience. The study also indicates that students and teachers of Ayurveda generally prefer Ayurvedic medicines as a first choice of treatment for all health needs of themselves and their family members. The study indicates that students and teachers of Ayurveda try to follow the directives given in *'Svasthavrtta'* under the topics like *'Viruddhāhāra'*, *'Sadvrtta'*, *'Dinacarya'* and *'Rtucharya'* in day to day living. The participants in the present study generally consider themselves to be capable of preparing simple Ayurvedic medicines sufficient enough for dispensing at their clinics. These are all positive trends and need to be looked at with a positive perspective. This indicates that students and teachers have inclination towards *'Pancakarma'* procedures and also are willing to rely on Ayurveda for their personal health needs.

- Students in the present study marginally tend to agree that they rarely prefer *'Rasaushadhis'* for their family members and close friends because of safety concerns. But teachers tend to disagree with the statement. This observation indicates that some apprehension prevails among students regarding the use of *'Rasaushadhi's* because of safety concerns.

Conclusion:

- BAMS graduates are required to be trained more vigorously in basic clinical skills

- Job opportunities after the completion of BAMS course are scarce and appropriate steps are needed to be taken to elevate the standard of education.

- Curriculum of BAMS course needs to be restructured by incorporating newer technological innovations and more practical material in all spheres.

- Changes are needed to be incorporated in teaching methodology making it more clinically oriented than memory oriented.

- Problems related to Global Challenges being faced by the Ayurvedic system of medicine need to be addressed by governing bodies.

- Entrepreneurship /Business opportunities after the completion of BAMS course are scarce at present and by improving the standards of education, the students may become more competent to enter the field of entrepreneurship.

- Ideal system of medical education for India would be to have a single kind of medical degree of graduate level, where basic training in Indian Systems of Medicine is given along with the training in conventional (Allopathic) medicine.

- At personal level, Ayurveda is considered to be relevant by the teachers and students who participated in the study.

Limitations of the present study and suggestions for future studies:

- The study does not cover the perceptions of Ayurvedic practitioners; rather, it represents the perceptions of students and teachers of Ayurveda only. The study therefore, only indicates the perception of academic sector. A similar study is suggested, which covers Ayurvedic practitioners also.

- Because of random sampling, the quality of education institutions was not considered in the study. If the similar study is carried out in only standard educational institutions, there are chances that results may vary.

- Only basic and general information regarding the teaching methodology and curricula of individual subjects can be had from the current study. Detailed studies in individual subjects are suggested to be carried out to understand the practical nature of problems involved in each subject.

- Out of 32 institutions covered in the study, 23 were of Governmental administration. Only 6 private and 3 semi-Governmental institutions were covered in the study. Also, maximum participants were from Governmental institutions (781). Therefore, this study reflects the status of Ayurvedic education system more in Governmental institutions than in private or semi-Governmental institutions. A similar study is suggested, where equal representation from all types of institutions may be ensured.

List of related Publications:

1. Patwardhan K, Gehlot S, Singh G, Rathore HC. The ayurveda education in India: How well are the graduates exposed to basic clinical skills? Evid Based Complement Alternat Med 2011;2011:197391.
2. Patwardhan K, Gehlot S, Singh G, Rathore HC. Global challenges of graduate level Ayurvedic education: A survey. Int J Ayurveda Res 2010;1:49-54.
3. Patwardhan K, Gehlot S, Singh G, Rathore HC. Graduate level Ayurveda education: Relevance of curriculum and teaching methodology. J Ayurveda 2009;3:74-82.
4. Patwardhan K, Gehlot S and Rathore HCS. Problems of Graduate Level Ayurvedic Education in India. University News. 2009;47[49]:07-13.
5. Patwardhan K. How practical are the "teaching reforms" without "curricular reforms"? J Ayurveda Integr Med 2010;1:174-6
6. Patwardhan K. Governance of higher education in Indian systems of medicine: Issues, concerns, and challenges. In: Kadam S, editor. Perspectives on Governance of Higher Education (Issues, concerns and challenges). Pune: Bharti Vidyapeeth Deemed University and Centre for Social Research and Development; 2010. p. 127-39.
7. Joshi H, Singh G, Patwardhan K. Ayurveda education: Evaluating the integrative approaches of teaching Kriya Sharira (Ayurveda physiology). J Ayurveda Integr Med [serial online] 2013 [cited 2013 Sep 26];4:138-46. Available from: http://www.jaim.in/text.asp?2013/4/3/138/118683

Bibliography and References

Annual Report on ISM&H 1999-2000. Department of Indian Systems of Medicine and Homeopathy, Government of India.

Antia NH (1981), **Report on Health for all: an alternative strategy,** New Delhi. Joint Panel of Indian Councils of Social and Medical Sciences Research, Govt. of India, 1981.

Astin JA (1998), **Why patients use alternative medicine: results of a national study,** *JAMA* 1998;279:1548-53.

AYUSH in India: 2007. **Introduction.** 2007;1-18. Available at http://indianmedicine.nic.in/Introduction.pdf

AYUSH in India: 2007. **Medical Education.** Section-4. 2007;83-144 available at http://indianmedicine.nic.in/Section%204.pdf

AYUSH in India: 2007. **Summary of All– India AYUSH infrastructure facilities.** Section-1. 2007;19-26 available at http://indianmedicine.nic.in/Section%201.pdf

Baghel MS (2006), **Issues in Publication of Ayurvedic Research Work – National & International Scenario – Shortcomings & Solutions,** http://www.serveveda.org/PDF%20Files/MS%20%20bagel.pdf accessed on 24th November, 2008 at 6:12PM

Bannerman RH (1983), **Traditional medicine and healthcare coverage.** Geneva: World Health Organization, 1983.

Bhardawaj PK and Upadhyay SD (2008), **Teaching of Ayurveda in Present Era,** Yogakshema Pravahika, Golden Jubilee issue on Ayurvedic Education, Editor: Acharya Pushpa Mitra (2008).p.21-22

Bhatta CP (2007), **Holistic Personality Development through Education: Ancient Indian Cultural Experiences,** Paper presented at the International Cultural Research Network and University of Strathclyde conference on "Exploring Cultural Perspectives in Education" held during May 3-6, 2007 at Glasgow, Scotland.

Bland JM and Altman DG. **Statistics notes: Cronbach's alpha.** *BMJ* 1997; 314;572

Bland JM and Altman DG. Statistics notes: Cronbach's alpha. BMJ 1997; 314;572

Board of Science and Education (2000), British Medical Association, **Acupuncture: efficacy, safety and practice,** Amsterdam: Harwood Academic, 2000.

Bodeker G (1999), **Traditional (i.e. indigenous) and complementary medicine in the Commonwealth: new partnerships planned with the formal health sector.** *J Alternative Complement Med* 1999; 5:97-101.

Bodeker G (2001), **Lessons on integration from the developing world's experience,** *BMJ* 2001;322:164–7

Caraka (700 BC), **Caraka Samhita,** English Translation. Sharma PV ed. Varanasi, India, Chaukhamba Orientalia.

Caspi O, Bell IR, Rychener D, Gaudet TW and Weil AT (2000), **The tower of Babel: communication and medicine. An essay on medical education and complementary-alternative medicine.** *Arch Intern Med* 2000;160:3193-5.

Chaudhari RR (2005), **'Research and evaluation of Traditional Medicine',** Paper distributed at SEAR Regional Meeting at Pyongyang, DPR Korea, 22-24 June 2005, quoted by M. S. Bghel, in 'Issues in Publication of Ayurvedic Research Work – National & International

Scenario – Shortcomings & Solutions' http://www.serveveda.org/PDF%20Files/MS%20%20bagel.pdf accessed on 24th November, 2008 at 6:12PM.

Chaudhury RR (1992), **'Herbal Medicine for Human Health'**. Regional Publication, 'SEARO' No. 20, WHO New Delhi.

Chi C (1994), **Integrating traditional medicine into modern health care systems: examining the role of Chinese medicine in Taiwan.** *Soc Sci Med* 1994;39:307-21

Cho HJ (2000), **Traditional medicine, professional monopoly and structural interests: a Korean case.** *Soc Sci Med* 2000;50:123-35.

Cooper EL (2008), **Ayurveda is Embraced by eCAM.** *Evid Based Complement Altern Med.* 2008;5(1)1–2.

Dash B and Kashyap L (1981 to 1982), **'Diagnosis and Treatment of Diseases in Ayurveda'**. In 5 Vols. Concept Publications Company, New Delhi.

Department of AYUSH **Development of Education system.** http://indianmedicine.nic.in/html/edu/aemain.htm#dev Data accessed on 4-10-2007, 6:30 PM.

Department of Indian Systems of Medicine and Homeopathy, Government of India. Annual report 1999-2000.

Department of Indian Systems of Medicines and Homoeopathy (2000). **Annual Report 1999-2000.** http://mohfw.nic.in/ismh/ (Data accessed 25 October, 2000.)

Dwarakanath C (1956), **Fundamental principles of Ayurveda**, Chaukhamba Surbharati Publications, 1956 (1999 reprint), Varanasi.

Dwivedi M and Gupta SN (2008), **Perspective of Ayurveda Education in India,** Yogakshema Pravahika, Golden Jubilee issue on Ayurvedic Education, Editor: Acharya Pushpa Mitra (2008).p.18-20

Easthope G, Beilby JJ, Gill GF, Tranter BK (1998), **Acupuncture in Australian general practice: practitioner characteristics.** *Med J Aust* 1998; 169: 197-200.

Frawely David (1989), **Ayurvedic Healing,** Motilal Banarasi Das Publications, New Delhi.

Gehlot S and Singh BM (2008), **Present ayurvedic education system and measures for improvement,** Yogakshema Pravahika, Golden Jubilee issue on Ayurvedic Education, Editor: Acharya Pushpa Mitra (2008).p.45-47.

Gogtay NJ, Bhatt HA, Dalvi SS, Kshirsagar NA. **The use and safety of non-allopathic Indian medicines.** *Drug Safety.* 2002;25(14):1005–19

Hesketh T, Zhu WX (1997), **Health in China. Traditional Chinese medicine: one country, two systems.** *BMJ* 1997; 315:115-7.

Hesketh TM, Zhu WX (1994), **Excessive expenditure of income on treatments in developing countries.** *BMJ* 1994;309:1441.

Indian Medicine Central Council (Minimum Standards of Education in Indian Medicine) (Amendment) Regulations, 1989. Available at http://www.ccimindia.org/Curriculum_ayurveda_1.htm

J Martin Bland and Douglas G Altman (1997), **Statistics notes: Cronbach's alpha**, *BMJ* 1997; 314;572

Joos S, Musselmann B and Szecsenyi J. **Integration of Complementary and Alternative Medicine into Family Practices in Germany: Results of a National Survey.** *Evid Based Complement Altern Med.* Advance Access published on March 17,2009; doi:10.1093/ecam/nep019

Joshi VK (2003) **Challenges in Ayurveda,** paper published in the souvenir of National Conference on Recent Advances in Ayurvedic Medicine, RAAM-2003, held at BHU on March 5-6, 2003. Chief Editor: V.K.Joshi. Published by: Faculty of Ayurveda, IMS, BHU.

Karnick CR (1994), **Pharmacdpeial Standard of Herbal Plants** in 3 Volumes, Sri Satguru Publications, New Delhi.

Kessler C (2006), **Criteria for the Establishment of Ayurveda** Hanover Medical School, Germany (Personal Communication at conference in Birstien, Germany). Quoted by M. S. Bghel, in 'Issues in Publication of Ayurvedic Research Work – National & International Scenario–Shortcomings & Solutions' http: //www.serveveda.org/ PDF%20Files/MS%20%20 bagel.pdf accessed on 24th November, 2008 at 6:12PM

Korotkov K. Book Review. The Scientific Basis of Integrative Medicine. *Evid Based Complement Altern Med.* 2005;2(3)425–426

Kumar S (2000), **India's Government promotes traditional healing practices.** *Lancet* 2000;335;1252.

Markandan N. (2001), **The Value and Purpose of Education,** *Contributions Towards An Agenda For India*, Indian Institute of Advanced Study, Shimla. Quoted by C. Panduranga Bhatta, **'Holistic Personality Development through Education: Ancient Indian Cultural Experiences',** Paper presented at the International Cultural Research Network and University of Strathclyde conference on "Exploring Cultural Perspectives in Education" held during May 3-6, 2007 at Glasgow, Scotland.

Mookerje RK (1989), **Ancient Indian Education**, New Delhi, Motilal Banarsidass. (Reprint). Quoted by Bhatta CP, **'Holistic Personality Development through Education: Ancient Indian Cultural Experiences',** Paper presented at the International Cultural Research Network and University of Strathclyde conference on "Exploring Cultural Perspectives in Education" held during May 3-6, 2007 at Glasgow, Scotland.

Morgan D, Glanville H, Mars S, Nathanson V, (1998), **Education and training in complementary and alternative medicine: a postal survey of UK universities, medical schools and faculties of nurse education.** *Complementary Ther Med* 1998;6: 64-70.

Muralidhara N (2008), **Change in Ayurveda education is the need of the hour,** Yogakshema Pravahika, Golden Jubilee issue on Ayurvedic Education, Editor: Acharya Pushpa Mitra (2008).p.38

National Health Interview Survey, USA, 2007, Available at: http://nccam.nih.gov/health/ayurveda/introduction.htm#ususe (Last Accessed on 2009 August 14).

National Policy on ISM&H (2002). Government of India. Available at http://www.whoindia.org/LinkFiles/AYUSH_NPolicy-ISM&H-Homeopathy.pdf (Last accessed on 2009 August 10).

News Report. **Export of Ayurvedic medicines affected due to high heavy metal content.** Available at: http://www.bio-medicine.org/medicine-news/Export-Of-Ayurvedic-Medicines-Affected-Due-To-High-Heavy-Metal-Content-6365-1/. (Last Accessed on: 2009 August 13).

Nuzzi R. Book Review. **Non Conventional Medicine in Italy. History, Problems, Prospects for Integration.** *Evid Based Complement Altern Med.* Advance Access Published on January 8, 2008; doi:10.1093/ecam/nem174

Osuide GE (1999), **Regulation of herbal medicines in Nigeria: the role of the National Agency for Food and Drug Administration and Control (NAFDAC).** Paper presented at the international conference on ethnomedicine and drug discovery. Silver Spring MD, Nov 3-5, 1999.

Pandey G (editor). The Caraka Samhita of Agnivesha (5th ed.). Varanasi, India: Chaukhambha Sanskrit Sansthan, vols. 1 and 2, The Kashi Sanskrit series no. 194.

Patwardhan B, Vaidya ADB and Chorghade M (2004), **Ayurveda and natural products drug discovery.** Current Science. 2004; 86(6):789-99

Patwardhan B, Warude D, Pushpangadan P and Bhatt N. **Ayurveda and traditional Chinese medicine: A comparative overview.** *Evid Based Complement Altern Med.* 2005;2(4)465-473

Patwardhan K and Gehlot S (2008), **Undergraduate Level Ayurvedic Education in India: Problems and Possible Solutions,** Yogakshema Pravahika, Golden Jubilee issue on Ayurvedic Education, Editor: Acharya Pushpa Mitra (2008).p.63-66

Planning Commission, Government of India, Eleventh Five Year Plan for Health and Family Welfare and AYUSH, Available at http://planningcommission.nic.in/plans/planrel/fiveyr/11th/11_v2/11v2_ch3.pdf

Ramajois M (1987), **National Unity, Equality, Rule of Law and Creating Men of Quality,** Dharwad, Karnatak University, Quoted by Bhatta CP, **'Holistic Personality Development through Education: Ancient Indian Cultural Experiences',** Paper presented at the International Cultural Research Network and University of Strathclyde conference on "Exploring Cultural Perspectives in Education" held during May 3-6, 2007 at Glasgow, Scotland.

Rampes H, Sharples F, Maragh S, Fisher P (1997), **Introducing complementary medicine into the medical curriculum.** *J R Soc Med* 1997; 90:19-22.

Rangachar S (1964) **Early Indian Thought,** Mysore, Geetha Book House, quoted by Bhatta CP, **'Holistic Personality Development through Education: Ancient Indian Cultural Experiences',** Paper presented at the International Cultural Research Network and University of Strathclyde conference on "Exploring Cultural Perspectives in Education" held during May 3-6, 2007 at Glasgow, Scotland.

Rao SKR (1985), **Encyclopedia of Indian Medicine** in 3 Volumes, Bombay, India, Popular Prakashan 1985.

Rastogi RP and Mehrotra RN (1993), **Compendium of Indian Medicinal Plants** in 3 Vol.New Delhi, India, CDRI, Lucknow and Publication and Information Directorate, 1993.

Saper RB, Kales SN, Paquin J, Burns MJ, Eisenberg DM, Davis RB and Phillips RS (2004), **Heavy metal content of ayurvedic herbal medicine products.** JAMA, 2004 December 15; 292 (23):2868-73.

Saper RB, Phillips RS, Sehgal A, Khouri N, Davis RB, Paquin J, Thuppil V and Kales SN (2008). **Lead, Mercury, and Arsenic in US- and Indian-Manufactured Ayurvedic Medicines Sold via the Internet.** JAMA. 2008;300(8):915-23.

Sharma BN (2008), **Present scenario of Ayurvedic education and remedies to improvise it to bring it at par with all different scientific educational system,** Yogakshema Pravahika, Golden Jubilee issue on Ayurvedic Education, Editor: Acharya Pushpa Mitra (2008). p.30-31

Sharma R.K. and Dash Bhagwan (2006), **Caraka Samhita,** Reprint Edition. Text with faithful and simple English translation and critical exposition based on Cakrapani Datta's Ayurveda Dipika. Chowkhamba Sanskrit Series Office. Varanasi. India.

Sharma PV (1972), **Indian Medicine in Classical Age**, Chaukhamba Publications, Varanasi, India.

Sharma PV (1994), **History of Medicine in India**, New Delhi, India, Indian National Science Academy, 1994.

Singh RH (1992), **Preventive and Social Medicine in Ayurveda in History of Medicine in India**: Indian National Science Academy, New Delhi.

Singh RH (1998), **The Holistic Principles of Ayurvedic Medicine**, Varanasi, India. Chaukhamba Surbharati Prakashan, 1998.

Singh RH (2000), **Ayurveda in India Today**. Procedures, International Symposium on Traditional Medicine, WHO Kobe Centre, Japan.

Singh RH (2000), **The unique holistic principles of Ayurvedic Diagnostic and cure.** Procedures, World Sanskrit Conference p. 123, Ministry of Human Resource Development New Delhi.

Singh RH (2001), **Basic principles of Ayurvedic Medicine and its materia medica.** In the Report on "Changing the Indian Health System -Current Issues and Future Directions" by Rajiv Mishra et al. Indian Council for Research on International Economic Relations New Delhi.

Singh RH (2001), Kayachikitsa (Internal Medicine), In **History of Science Philosophy and Culture in Indian Civilisation** Vol. IV Part 2, Ch. 4 pp. 128-156. Ed. RV. Subbarayappa, Centre of Studies in Civilisation, New Delhi.

Singh RH (2003), **Experiences With Alternative Systems of Medicine In India** paper published in the souvenir of National Conference on Recent Advances in Ayurvedic Medicine, RAAM-2003, held at BHU on March 5-6, 2003. Chief Editor: V.K.Joshi. Published by: Faculty of Ayurveda, IMS, BHU.

Sridhar N (2008), **Challenges Before Ayurvedic Education – Solutions** http://www.serveveda.org/PDF%20Files/Dr.%20N.%20Sridhar.pdf accessed on 24th November, 2008 at 6:09PM

Srinivasulu M (2006), **Challenges Before Ayurvedic Education, Practice-Solutions** http://www.serveveda.org/PDF%20Files/Dr.%20M.%20Srinivasulu.pdf (last accessed on 24th November 2008 at 6:20 PM.)

Streiner D and Norman G (1995), Health Measurement Scales: A Practical Guide to Their Use. Oxford: Oxford University Press; 1995.

Streiner D, Norman G. Health Measurement Scales: **A Practical Guide to Their Development and Use.** 2nd Edition. Oxford, Oxford University Press. 1995.

Stumpf SH and Shapiro SJ. **Bilateral Integrative Medicine, Obviously**. *Evid Based Complement Altern Med.* 2006;3(2)279–282

Sundberg T, Halpin J, Warenmark A and Falkenberg T. **Towards a model for integrative medicine in Swedish primary care**. *BMC Health Services Research.* 2007; 7:107 doi:10.1186/1472-6963-7-107

Susruta, **Susruta Samhita** (700 BC), Singhal *G.D.* et al. eds. Varanasi, India, Chaukhamba Surbharati.

Svoboda RE. **Ayurveda - Life, Health and Longevity**. The Ayurvedic Press, Albuquerque 87112, New Mexico. 2004; ISBN 1883725097

Taittiriya Upanishad. **Ten Principal Upanishads,** Vol 1 (1987), New Delhi, Motilal Banarsidass. Quoted by Bhatta CP, **'Holistic Personality Development through Education: Ancient Indian Cultural Experiences',** Paper presented at the International Cultural Research Network and University of Strathclyde conference on "Exploring Cultural Perspectives in Education" held during May 3-6, 2007 at Glasgow, Scotland.

Tewari PV (2008), *Ayurved Ki Shiksha,* Yogakshema Pravahika, Golden Jubilee issue on Ayurvedic Education, Editor: Acharya Pushpa Mitra (2008) p.7-8

Thatte U, Bhalerao S. (2008), **Pharmacovigilance of ayurvedic medicines in India**. Indian J Pharmacol 2008;40:10-12

Thatte UM, Rege NN, Phatak S, Dahanukar SA. **The flip side of ayurveda?** J Postgrad Med

1993;39:179-82.

The gazette of India (2005), Extraordinary, Part-III, Section-4, No.17, New Delhi, Dated 4-2-2005

The State Administration of Traditional Chinese Medicine of the People's Republic of China. **Anthology of policies, laws and regulations of the People's Republic of China on traditional Chinese medicine.** Shangdong: Shangdong University, 1997.

Udupa KN (1958), **The Udupa Committee report on indigenous systems of medicine in India**, New Delhi, India, Ministry of Health, 1958 Publication.

Udupa KN (1974), **Advances in research in Indian Medicine**, Varansi, India, Banaras Hindu University Publication Cell, 1974.

Udupa KN, Singh RH (1970), **Science and philosophy of Indian medicine**, Nagpur, India, Shri Baidyanath Ayurveda Bhawan, 1970.

Vickers A (2000), **Recent advances: Complementary medicine.** *BMJ* 2000;321:683-6. (16 September.)

What does Cronbach's alpha mean? Available at: http://www.ats.ucla.edu/stat/Spss/fa%E2%80%A6%00%_00 accessed on 25-11-2008 at 7:24 PM.

WHO (1978), **Health for All by 2000 A.D**. Alma Ata Declaration. World Health Organisation, Geneva.

WHO (2000), **General guidelines for methodologies on research & evaluation of Traditional Medicine**, WHO/EDM/TRM/ 2000.1, GENEVA quoted by M. S. Bghel, in 'Issues in Publication of Ayurvedic Research Work – National & International Scenario – Shortcomings &Solutions'http://www.serveveda.org/PDF%20Files/MS%20%20bagel.pdf accessed on 24th November, 2008 at 6:12PM

Wujastyk D and Smith FM. (2008), Editors. **Modern and Global Ayurveda, pluralism and paradigms.** State University of New York Press (Publisher). ISBN: 9780791474891.

Zollman C and Vickers A (1999), **'ABC of complementary medicine: What is complementary medicine?'** *BMJ* 1999; 319;693-696

Zysk KG (1996), **Medicine in Veda**. New Delhi, India, Motilal Banarasi Das, 1996.

Annexure-1 (Questionnaire)

'Relevance of Current System of Ayurvedic Education in the Emerging Global Scenario'

Name: …………………………..Male ☐ Female ☐ Age………..……
Institution:……………………………..…………………………………………
Present Status: Teacher ☐ Student ☐
a) If a teacher : **Designation**…… **Qualification**………**Subject**………………
b) If a student : Enrolled under: B.A.M.S. ☐ M.D.(Ay) ☐ M.S.(Ay) ☐ Ph.D ☐

1. Please go through the following list of statements related to the "Extent of exposure to the basic clinical skills during BAMS Course" and indicate your level of agreement with each statement in terms of 'Strongly Agree'(SA), 'Agree'(A), 'Undecided'(U), 'Disagree'(D) and 'Strongly Disagree'(SD).

	During the B.A.M.S. course:	SA	A	U	D	SD
1	Students are not trained to handle the clinical emergencies of primary healthcare level, through Ayurvedic methods.					
2	Students are not exposed to any successful Ayurvedic method of primary healthcare in the management of infectious conditions like malaria and tuberculosis.					
3	Students are not exposed to any successful Ayurvedic method of primary healthcare in the management of poisoning.					
4	Students are not exposed sufficiently to the basic clinical skills and procedures like incision and drainage, suturing and catheterization.					
5	Students are not trained sufficiently to conduct normal delivery.					
6	Students are not exposed to a large variety of cases because patients visiting Ayurvedic institutions belong to only few identifiable categories like those complaining of joint pain, ano-rectal diseases, stroke etc.					
7	Students are not exposed sufficiently to the basic modern knowledge of the subjects like Physiology, Pathology, Biochemistry, Pharmacology, Medicine, Pediatrics, Obstetrics & Gynaecology, Eye & ENT and Surgery.					
8	Students are not exposed sufficiently to the basic skills of interpreting ECG, X-Ray and such other diagnostic tools and their clinical utility.					
9	Students are not trained in the basic skills in the areas like Genetic counseling, Human sexuality, End of life care, Geriatrics and Drug and alcohol abuse.					
10	Students are not trained sufficiently in the basic clinical methods related to *Panca Karma*, *Kshara Sutra* and *Jalaukavacharana*.					
11	Students are not exposed sufficiently to the basic methods of physical examination, diagnosis and management of common clinical conditions, making them non-confident clinicians/ practitioners.					

2. **Please go through the following list of statements related to the "Job opportunities after the completion of BAMS course" and indicate your level of agreement with each statement in terms of 'Strongly Agree'(SA), 'Agree'(A), 'Undecided'(U), 'Disagree'(D) and 'Strongly Disagree'(SD).**

		SA	A	U	D	SD
1	Legally, in most of the states, a BAMS degree holder cannot practice Allopathy and therefore hospitals generally prefer MBBS graduates as medical officers instead of BAMS graduates.					
2	Ayurvedic hospitals are less in number in comparison to Allopathic ones and therefore job opportunities are limited.					
3	In Ayurvedic educational institutions, only Post Graduate doctors are employed and not BAMS degree holders.					
4	Most of the research institutions prefer Post Graduate doctors and therefore, job opportunities in research institutions are limited.					
5	Even in Government sector, BAMS graduates are not treated at par with MBBS graduates and therefore, job opportunities are limited in certain areas e.g., Railways.					
6	Ayurvedic pharmaceutical firms prefer Post Graduate candidates to BAMS degree holders as experts.					
7	There is lot of competition for jobs among BAMS degree holders as a result of mushrooming of Ayurvedic colleges.					

3. **Please go through the following list of statements related to the "Scientific relevance of the Curriculum of BAMS course" and indicate your level of agreement with each statement in terms of 'Strongly Agree'(SA), 'Agree'(A), 'Undecided'(U), 'Disagree'(D) and 'Strongly Disagree'(SD).**

		SA	A	U	D	SD
1	Most of the topics in the subject *'Ayurvediya Itihasa'* have got least practical applicability.					
2	Most of the topics covered in the subject *'Padartha Vijnyana'* are philosophical and their practical applicability is limited.					
3	Many topics in *'Rachana Sharira'* like *'Marma'*, *'Sira'*, *'Snayu'*, *'Sandhi'* etc. are outdated as more advanced knowledge on these topics is available in the textbooks of Modern Anatomy/ Modern surgery.					
4	Topics like 'Assessment of *Prakriti* and *Dhatu Sara*' are given undue importance in the subject *'Kriya Sharira'* and the clinical applicability of these topics is not emphasized in clinical disciplines.					
5	The essential practical exposure to the laboratory diagnostic methods in serology, immunology, histopathology, microbiology and parasitology is not					

	emphasized in *'Roga nidana* and *Vikriti Vijnana'*.					
6	In *'Dravyaguna'*, essential basic information related to recent advances in pharmacodynamic/ pharmacognostic/ phytochemical attributes of various Ayurvedic herbs and methods of evaluation of their pharmacological effects is not emphasized.					
7	Essential basic knowledge related to various technologically advanced methods of 'Drug Standardization' is not included in the curricula of either *'Dravyaguna'* or *'Rasa Shastra'*.					
8	Essential basic knowledge related to pharmaco-vigilance, safety profile, toxicity studies and Good Manufacturing Practices– is not included in *'Rasashastra'*.					
9	Essential basic knowledge related to the methods of quantitative and qualitative analysis of chemical components of Ayurvedic preparations is not included in the curriculum of *'Rasashastra'*.					
10	In *'Agada Tantra'*, most of the Ayurvedic topics describing the classifications/numbers/varieties of poisons and their effects are outdated and impractical.					
11	Topics related to *'Arishta Vijnana'* explained in *'Indriya Sthana'* of *'Caraka Samhita'* are practically not useful because they do not fit in to the present social scenario.					
12	The detailed explanations related to the preparation / measurements of instruments used in *'Panchakarma'* (e.g., *'Basti Netra'*, *'Basti Putaka'*) / their defects / complications of wrong use etc. are not practically useful and therefore, are not relevant.					
13	Practical training related to the basics of medical jurisprudence, toxicology and forensic medicine is not emphasized in teaching making a BAMS graduate inefficient in handling the legal procedures.					
14	Essential information of recent studies/ reports related to efficacy of Ayurvedic medicines/ procedures is not included in the curriculum of clinical disciplines, which is required to be included.					
15	The curricula of clinical disciplines contain many outdated methods of treatment/management which are impractical (e.g., *Droni Praveshika Rasayana*).					
16	Certain essential topics like Clinical decision making, Care of hospitalized patients, Patient interviewing skills, Ethical decision making and Geriatric patient care are not emphasized in the curricula of the clinical disciplines, which are required to be included.					

17	Certain essential topics related to medical practice including Cost effective medical practice, Quality assurance in medicine, Practice management and Medical record-keeping are not included in the curriculum, which are required to be included.					
18	Many modern technical terms are translated into 'Sanskrit' in the curriculum (e.g., *'Unduka Puccha Shotha'* for Appendicitis, *'Hritkaryacakra'* for cardiac cycle etc.) which does not serve any practical purpose.					
19	Many controversial topics (e.g., certain structures in *Rachana Sharira*, certain herbs in *Dravyaguna*) are included in the curricula which lead only to confusion among students.					

4. **Please go through the following list of statements related to "Teaching methodology that is followed in existing system of Ayurvedic education" and indicate your level of agreement with each statement in terms of 'Strongly Agree'(SA), 'Agree'(A), 'Undecided'(U), 'Disagree'(D) and 'Strongly Disagree'(SD).**

		SA	A	U	D	SD
1	Ayurvedic teaching methodology does not keep up the scientific values and scientific spirit of a young student.					
2	Ayurvedic teaching methodology does not encourage questioning among the young students.					
3	Interpretation of theories like *'Tridosha'* or *'Pancha Mahabhuta'* varies largely from one teacher to another making them further confusing and vague.					
4	Memorizing the classical 'Sanskrit' verses is unduly emphasized in Ayurvedic method of teaching, making the process of learning difficult.					
5	Memorizing the reference number of a particular chapter/ verse of any *'Samhita'* does not serve any practical purpose, but is given undue importance in teaching and examination system.					
6	The examination system in Ayurveda does not assess the actual abilities and skills of a student; rather, it largely depends on the assessment of memorizing capacity of students.					
7	Standard textbooks covering all the topics in curricula are not available in the market, making understanding on the subjects more difficult.					
8	'Sanskrit' is given undue importance in teaching – learning process making the process of understanding the subject difficult.					
9	Memorizing the numbers of various structures / their measurements / various classifications of diseases *(Sankhya*					

	Samprapti) as per different authors etc. does not serve any practical purpose, but is given undue importance in teaching and examination system.					
10	Students are not trained sufficiently in certain areas of 'Evidence based medicine' like Interpretation of clinical data and research reports, Literature reviews/critiques, Interpretation of laboratory results and Decision analysis.					
11	Students are not trained in areas like using a computer-based clinical record keeping program, carrying out reasonably sophisticated searches of medical information databases on internet, using a variety of forms of telemedicine etc., making them technologically inferior.					
12	Students are not introduced to essential 'Communication skills' like discussing a prescription error with the patient, providing safe sex counseling, negotiating with a patient who requests unnecessary investigations etc.					

5. **Please go through the following list of statements related to "Global Challenges being faced by the Ayurvedic syestem of medicine" and indicate your level of agreement with each statement in terms of 'Strongly Agree'(SA), 'Agree'(A), 'Undecided'(U), 'Disagree' (D) and 'Strongly Disagree'(SD).**

6.

		SA	A	U	D	SD
1	Serious questions are being raised on the safety profile of Ayurvedic preparations in some countries posing a threat to the Ayurvedic system of Medicine.					
2	Standardization of Ayurvedic preparations is still a problem that needs to be addressed.					
3	In many countries, legally, practicing Ayurveda is not allowed and therefore, there are no opportunities for BAMS graduates in such countries.					
4	Possible entry of foreign universities in India may pose a threat to the existing educational institutions.					
5	Ayurvedic academicians do not figure anywhere in authoring the scientific and evidence based papers in reputed international journals.					
6	Ayurvedic academicians do not voluntarily participate in International platforms to present their research data.					
7	Ayurvedic academicians do not follow international standards while planning the protocols of research projects and while writing research reports.					
8	Ayurvedic scholars generally do not have knowledge regarding 'Intellectual Property Rights' and patenting procedures.					
9	Authentic websites providing up-to-date knowledge in Ayurveda are not hosted by Ayurvedic institutions.					
10	No standard international indexed and peer-reviewed journals are published by Ayurvedic institutions making it difficult for					

	Ayurvedic researches have global attention.					
11	Pharmacodynamic/ pharmacokinetic properties/ efficacy/ safety profiles and chemical compositions of Ayurvedic formulations are yet to be established making it difficult for experts in conventional medicine to accept Ayurveda.					

7. **Please go through the following list of statements related to "Entrepreneurship /Business opportunities after the completion of BAMS course" and indicate your level of agreement with each statement in terms of 'Strongly Agree'(SA), 'Agree'(A), 'Undecided'(U), 'Disagree'(D) and 'Strongly Disagree'(SD).**

		SA	A	U	D	SD
1	Students are not trained in management skills required to launch a new Ayurvedic hospital/ Panchakarma center/ Ayurvedic Pharmacy during BAMS course.					
2	Students are not exposed to the basics of economical aspects related to healthcare sector during BAMS course.					
3	Most of the BAMS graduates prefer either studying PG course or they go for private practice and therefore, inspiring examples of industrially successful BAMS graduates are very few.					
4	Students are not introduced to the skills related to the management of Health tourism and emerging opportunities in this field during BAMS course.					
5	Students are not exposed to the agricultural and marketing aspects of medicinal plants making it difficult to go for cultivation / marketing of medicinal plants.					
6	Students are not exposed to the manufacturing techniques related to cosmetic products and such other popular dosage forms during BAMS course making them unfit for modern pharmaceutical industry.					

7. In your opinion, the "Ideal system of medical education for India" would be:
   (Mark (√) against any one statement that you think is the best option).

1	The existing system of multiple streams of medical education should continue as such in parallel with conventional modern medical education.	
2	Only one kind of medical degree of graduate level has to be there in India. Subject content of Indian medical systems like Ayurveda, Unani, Siddha must be included in the same curriculum along with conventional modern medicine.	
3	Only one kind of medical degree of graduate level has to be there in India which should be in conventional modern medicine. Only at Post Graduate level, one should be given an option of alternative systems of medicine.	

8. Please go through the following list of statements related to "Personal relevance of Ayurveda" and indicate your level of agreement with each statement in terms of 'Strongly Agree'(SA), 'Agree'(A), 'Undecided'(U), 'Disagree'(D) and 'Strongly Disagree'(SD).

		SA	A	U	D	SD
1	I personally have not undergone any Panchakarma procedure during my student life to have a firsthand experience.					
2	I personally don't prefer Ayurvedic medicines as a first choice for all health needs of myself and my family members.					
3	I rarely prefer 'Rasaushadhis' for my family members and close friends because I am not convinced about their safety.					
4	I do not follow the directives given in Svasthavritta under the topics like 'Viruddhahara', 'Sadvritta', 'Dinacharya' and 'Ritucharya' in day to day living.					
5	I cannot precisely identify most of the herbs used in Ayurvedic pharmaceutics.					
6	I am not confident of preparing simple Ayurvedic medicines sufficient enough for dispensing in my clinical practice.					

Any other relevant observation / suggestion:

Declaration:

I hereby declare that my participation in the study titled **'Relevance of Current System of Ayurvedic Education in the Emerging Global Scenario'** is purely voluntary. Also, all the responses given are based on my own views and I was not compelled to respond in any particular way by the investigators or by any other authority.

Signature of the participant

www.ingramcontent.com/pod-product-compliance
Lightning Source LLC
Chambersburg PA
CBHW080930170526
45158CB00008B/2233